Industrial Gas Chromatographic Trace Analysis

Industrial Gas Chromatographic Trace Analysis

Horst Hachenberg

Analytical Chemistry Section, Farbwerke Hoechst A.G.
Frankfurt am Main, Federal Republic of Germany

London · New York · Rheine

Heyden & Son Ltd., Spectrum House, Alderton Crescent, London NW4 3XX.

Heyden & Son Inc., 225 Park Avenue, New York, N.Y. 10017, U.S.A.

Heyden & Son GmbH, 4440 Rheine/Westf., Münsterstrasse 22, Germany.

Translated by D. Verdin

Library of Congress Catalog Card No. 73–83338

ISBN 0 85501 079 7

Printed in Great Britain by Galliard (Printers) Ltd., Great Yarmouth, England

PREFACE

This book is concerned with the applications, potentialities and difficulties of gas chromatographic trace analysis in industry. This very specialized technique can only very briefly be included in courses on analytical chemistry at universities and technical colleges. Their primary consideration is to provide a broad education, and also the very expensive apparatus needed is generally not available for educational purposes. It is hoped that this book will serve as a guide to those who need to employ this analytical technique. Naturally, if a book of this scope is to be sufficiently informative it cannot encompass the whole field of gas chromatographic trace analysis. The following discourse is therefore based mainly on areas to which the author has himself contributed during many years of analytical research in the fields of petrochemistry and synthetic polymers. Even here such an extensive range of problems arises that the analytical chemist has difficulty in dealing with them in addition to his normal work. Consequently, new problems in trace analysis must be solved as quickly as possible, and developed into routine and, if possible, automated analytical procedures.

The work of the industrial analytical chemist is largely governed by economic and material objectives, and consequently often leaves too little opportunity for a detailed academic approach to the subject. Perhaps the reader will therefore bear with the author if some of the papers referred to are not discussed as extensively as he might consider necessary. This may appear unsatisfactory to those who have recently entered industry from technical colleges and universities. Nevertheless, they will soon recognize that this branch of physical analysis represents an important and often indispensable link between research and production, since gas chromatographic trace analysis assists in the development of new production techniques and in the improvement of existing methods. Moreover, the experience and knowledge gained frequently become part of the 'know-how' related to a process.

The author thanks Farbwerke Hoechst AG for permission to publish this book.

Frankfurt am Main.
August 1973.   H. Hachenberg

CONTENTS

PART 2

Applications of Trace Analysis

PART 1

General Principles

1.1. THE TERM 'TRACE ANALYSIS'

Trace analysis means the determination of materials which are present in very low concentrations in pure substances, or in mixtures of substances, and which are generally referred to as trace components.

The field of trace analysis is frequently characterized by defining the limits of detection. Although this may be justified in many individual cases, generally speaking the scope of trace analysis cannot be defined absolutely. The definition of the range of concentrations depends largely on the particular problem in relation to the analytical methods available, and thus in practice it is relative.

This is especially true in the case of gas chromatographic trace analysis in which separation and determination of the individual components are carried out in one operation. Thus, it is frequently the case that the limit of detection is specified on the basis of a calculated sensitivity for the detector, or on an experimentally determined value for a definite group of substances.

Apart from the fact that uncertainties arise in comparisons of sensitivities between gas chromatographic detectors, the result of a gas chromatographic trace determination ultimately depends on many other parameters. Thus, for example, the efficiency of separation and its dependence on sample size, as well as the stability of the stationary phases on which difficult separations are carried out, are equally important factors. In addition to this, as in other methods of trace analysis, the composition of the material, the nature and state of the sample, the method by which it is collected, the material from which the sample container is made, and the location of the sampling point, etc. can all play a decisive role. For these reasons some theoretical limits of detection, which have been defined solely on the basis of the sensitivity of the detectors employed, cannot be achieved in practice. Special attention should be paid to this when drawing up specifications for high purity materials, since the best specification is useless if it cannot be validated because of an inadequate analytical technique.

In practice, a further important question concerns the requirements from a trace analysis and the concentration level at which it is meaningful to determine the trace components. In general, trace analyses will be necessary when industrial processes are endangered by corrosion and safety risks. Other examples are the purity of monomers used in polymerization reactions where trace amounts of foreign substances can cause chain termination or branching, and the materials employed for catalytic processes where the active life of a catalyst is often determined by the trace constituents. Reference may also be made to the quality control of manufactured goods in which trace components can cause problems of discoloration, degradation or undesirable odours, or can give rise to health hazards. The analytical problems associated with atmospheric and water pollution should also be mentioned.

The term 'trace analysis' also depends on the requirements of the user and on the problem. Thus, for example, the determination of a component present at 100 ppm can be more important and far more difficult than that of another component present at 10 ppm or *vice versa*.

The analytical technique is itself undergoing continual development. Whereas 15 years ago a gas chromatographic analysis in the region of 0.1% to 0.01% was still considered to be very successful, it is now common, even in routine operation, to analyse in the range $10^{-4}\%$ to $10^{-6}\%$ and down to $10^{-7}\%$ or less in special cases.

The results of gas chromatographic trace analyses are therefore no longer given as percentage concentrations but in ppm (parts per million) and ppb (parts per billion), where

1 ppm by weight $= 10^{-4}$ weight $\%$ (weight/weight)
1 volume ppm $= 10^{-4}$ volume $\%$ (volume/volume)

and

1 ppb by weight $= 10^{-7}$ weight $\%$
1 volume ppb $= 10^{-7}$ volume $\%$

1.2. AIMS OF GAS CHROMATOGRAPHIC TRACE ANALYSIS

In trace analysis, three basically different aims may be distinguished.

(*a*) Quantitative analysis of trace components of known identity.

In this case, the material to be analysed and the method are known. In addition to optimizing the conditions required for resolution on the gas chromatographic column and the operating conditions of the detector, it is of major importance that the calibration be correctly carried out. Careful attention must also be given to the noise level of the whole instrument, including the precision of the signal recorder and the associated limits of detection. Nowadays this kind of analysis is performed even with commercial gas chromatographs.

(*b*) Detection, identification and quantitative determination of specific trace materials in pure substances and mixtures of known substances.

Here the objective is well defined. The analysis is therefore relatively simple since the conditions under which resolution will occur, the choice of the detector and the sampling technique may be assessed on the basis of known parameters such as the chemical structure, polarity and boiling point of the trace component to be analysed. Measurement of retention times ensures a high probability of accurate identification in this instance.

(*c*) Detection, identification and quantitative determination of unknown trace components in materials of largely or completely unknown composition.

In this case, there is initially no information on the material to be analysed. It may, for example, involve the analysis of mixtures of products from new processes in which corrosion may take place, or of manufactured products in which discoloration, degradation and irritating odours can be detected. There are also difficult analytical problems associated with polymerization processes and in the examination of commercial samples.

The first task confronting the analyst is to isolate all the trace components and quantitatively estimate them as far as possible. Except in special cases, this can only be achieved by gas chromatography since it enables separation and measurement to be performed in one operation. In order effectively to distinguish all the trace components, the analyst must employ several different polar columns in combination with universal and specific detectors. The use of a single column with only one detector is not sufficient and can lead to serious errors because important impurities can be overlooked owing to poor resolution or insensitive detection.

When it is considered that this problem has been solved, then identification of the trace constituents isolated must be confirmed since gas chromatography is generally a 'blind' technique. For materials present in trace amounts identification is the most difficult part of the analysis, and in spite of the many concentration techniques available, and also the capabilities of combined gas chromatography and mass spectrometry in the ppm region it can by no means always be achieved. Thus, in many cases it is not possible to completely analyse the trace components in unknown materials, but merely to indicate the presence of traces of unidentified compounds.

1.3. REQUIREMENTS FOR GAS CHROMATOGRAPHIC TRACE ANALYSIS

The first necessity for a gas chromatographic trace analysis is a sufficiently sensitive means of detecting components which have been separated. The method therefore depends primarily on the detector. If trace components are sufficiently well resolved from the other constituents, it is possible to increase the limit of detection by using larger samples. If this approach is not successful then there is little choice but to undertake time-consuming concentration procedures.

Before commencing a gas chromatographic trace analysis it is also necessary to consider the noise level of the whole instrument, as well as the best methods of sampling, sample injection and calibration. When this assessment has been made the crucial question of method of identification often still remains.

1.31. Detectors

The literature contains a large amount of information on detectors. Particular reference may be made to the publications of Lovelock, Gogh and Walker, and that of Halász in which an excellent survey and classification of the gas chromatographic detectors currently available is presented.[1-3] In the course of this book, for instructional purposes, detectors will be discussed in detail where they are applied to particular problems. The following sections therefore discuss only a few basic considerations from the author's own experience of the application of detectors to trace analysis.

Since the effort involved in trace analysis usually exceeds that for analyses in the percentage region, owing to the difficulty of calibration and the need for blank analyses, the linearity of the detector is not of such crucial importance as obtaining the highest possible selectivity and sensitivity with a noise level for the whole instrument as low as possible. The detector should also be largely insensitive to fluctuations in temperature, pressure and flow rate, and should be unaffected by contamination. Since in trace analysis the components to be determined are usually poorly resolved from the main constituents, the volume of the detector chamber should be small so as to minimize the amount of back-diffusion that can occur.

Excluding special cases, the following types of detector are employed for gas chromatographic trace analysis, which from a practical point of view may be characterized as below.

1.311. Non-specific detectors
THERMAL CONDUCTIVITY CELLS
Operated at low temperature and having the smallest possible detector volume.
Limit of detection: ~0·5 ppm for a gas sample size of 5 ml.
Response: Universal.
Applications: For materials to which the flame ionization detector (FID) does not respond; however, they are restricted to gases as a consequence of their low operating temperature.
Special considerations: Very sensitive to fluctuations in the temperature, flow rate and power supply.

HELIUM DETECTOR
Limit of detection: ~ 100–10 ppb for a gas sample size of 1 ml.
Response: Universal.
Applications: For all gases to which the FID does not respond, *i.e.* mainly inorganic and inert gases. An ideal supplement to the FID for the trace analysis of complex mixtures of gases.
Special considerations: The following difficulties result from the extremely high sensitivity: The quality of the gas sampling system must be of the highest standard. If it contains moving parts it must be provided with a helium purge to prevent the diffusing in of air. For the same reason, if traces of oxygen, nitrogen or carbon dioxide are to be measured, accurate sampling is possible only if special precautions are taken. A further difficult and time-consuming problem arises from the fact that the helium detector is just as sensitive to water vapour as to other substances. The water vapour taken in with each sample is concentrated in the column during the course of several analyses and after a certain time is eluted as a broad peak, causing the sensitivity of the detector to fall sharply and making further trace analysis impossible. Depending on the number of samples, their size, and their water vapour content, this problem occurs after about 6–8 hours, and in laboratory gas chromatographs it is overcome by prolonged heating. It is also necessary to keep the concentration of water vapour and other contaminants in the carrier gas as low as possible. The helium

employed is therefore purified by passing through a molecular sieve at liquid nitrogen temperature before it enters the gas chromatograph. Liquids cannot yet be analysed with this detector since, for the reasons already mentioned, normal sampling systems do not allow the use of a syringe injecting through a rubber septum. Unfortunately, only adsorption column packings such as activated charcoal, silica gel, aluminium oxide, molecular sieves or cross-linked polymers can be used since even a minute amount of column bleeding from a liquid stationary phase renders trace analysis impossible. Applications of the helium detector are therefore restricted to gases for the time being.

ARGON IONIZATION DETECTOR

Limit of detection: 100–500 ppb for a sample of 3–5 ml of gas or 0·01–0·02 ml of liquid.

Response: All substances having ionization energies below 11·6 eV.

Applications: Organic compounds. Trace amounts of inorganic and inert gases can only be determined under special operating conditions.

Special considerations: Sensitive to contaminants, water and column bleeding. The choice of stationary phase is therefore restricted.

RADIOFREQUENCY DISCHARGE DETECTOR

Limit of detection: ~100–500 ppb for a gas sample size of 5 ml.

Response: Universal.

Applications: Inorganic and inert gases, and gaseous saturated hydrocarbons. It is a good supplement to the FID for the trace analysis of complex mixtures of gases.

Special considerations: Sensitive to contaminants and column bleeding. Its applications are limited since polymerizable compounds easily form deposits on the walls of the detector and make it insensitive or temporarily unserviceable. It can therefore not be employed for the trace analysis of monomers. It is possible to destroy the polymeric film by having oxygen present.

1.312. Specific detectors

A distinction can be made between detectors which are specific for classes of substances and those which are specific for particular compounds. Strictly speaking only the mass spectrometer should be included among the latter.

FLAME IONIZATION DETECTOR (FID)

Limit of detection: ~ 10 ppb for a sample size of 2–3 ml of gas or 0·01–0·02 ml of liquid.

Response: All organic compounds except formaldehyde, formic acid and some perchlorinated hydrocarbons to which it responds with a very low sensitivity.

Applications: Universal use in the trace analysis of organic compounds which vaporize without decomposition. Trace analysis can be carried out on capillary columns if a good amplifier is used. Since it does not respond to inert gases and water, it is particularly suited to the analysis of traces of organic impurities in air and water.

Special considerations: Insensitive to fluctuations in temperature and carrier gas flow rate. It is therefore well suited to the routine analysis of trace amounts of materials. Sensitive to fluctuations in the flow rate and impurities in combustible gases: air (oxygen) and hydrogen. This is generally the most reliable detector for trace analysis.

ELECTRON CAPTURE DETECTOR (ECD)

Limit of detection: ppb region.

Response: Sensitive to substances having a high electron affinity such as halogen and nitro compounds, carbonyls and nitriles, organo-metallic compounds and substances with conjugated double bonds which contain aldehydic or ketonic groups within the conjugated system. No response to saturated hydrocarbons. Since the response is strongly dependent on the bond energy of the relevant hetero atom, very large differences in sensitivity occur between individual classes of compounds.

Applications: Specific detection of traces of substances in the groups mentioned above when present in mixtures of hydrocarbons or in air.

Special considerations: Sensitive to fluctuations in temperature, impurities in the carrier gas and from the column, and also to water when a tritium radiation source is used. The sensitivity under varying analytical conditions is often not particularly constant, nevertheless it is adequate for routine measurements. However, frequent calibration is necessary.

HALOGEN–PHOSPHORUS DETECTOR (HPD)

(Also called the alkali flame ionization detector (AFID) or the thermionic detector (TID).)

Limit of detection: This is generally based on parathion, of which 2×10^{-10} g can be detected. The limit of detection for halogen compounds can be given as 0·5–1 ppm for a sample size of 5 ml of gas or 0·02 ml of liquid, whilst that for phosphorus compounds is generally lower.

Applications: Specific detection of halogen and phosphorus compounds.

Special considerations: Very good specificity and reliable operation. However, some commercial models show a fall in sensitivity when first brought into use and only settle down after a number of hours.

COULOMETRIC DETECTOR (CMD)

Limit of detection: 1 ppm.

Applications: Specific detection of traces of sulphur and halogen compounds in mixtures of substances, particularly hydrocarbons.

Special considerations: Relatively difficult to operate since chemical reactions can take place before detection. It can only be used in the laboratory owing to the risk of breakage.

FLAME PHOTOMETRIC DETECTOR (FPD)

Limit of detection: ~ 100 ppb for a sample of about 5 ml of gas or about 0·03 ml of liquid.

Applications: Specific detection of sulphur and phosphorus in mixtures of substances.

Special considerations: Good stability, simple operation. Complications arise owing to quenching effects if trace components are masked by those present in concentrations at the percentage level.

ATOMIC ABSORPTION (AAS)

Limit of detection: ppb region.

Applications: metallic compounds in gases and liquids.

MASS SPECTROMETER

Limit of detection: When used in combination with a direct ion source and suitable capillary columns, about 10 ppm (measured as hydrocarbons) gives a mass spectrum which can still clearly be interpreted.

Response: Universal.

Applications: All substances which can be eluted by gas chromatography.

Special considerations: An efficient differential pumping system is required for trace analysis. Interpretation at the ppm level becomes difficult, or impossible, if the component concerned does not show a peak for the parent molecule.

Hartmann[4] has clearly summarized the most important characteristics of detectors and pointed out that they may be divided into two groups. Those in the first group measure a composite property of the sample and the carrier gas, the measured value being related to the concentration of the sample in the carrier gas. Dimbat et al.[5] defined a unit of sensitivity corresponding to the sensitivity of a detector which gives an output of 1 mV when the sample concentration is 1 mg of vapour/ml of carrier gas. Hence, the sensitivity is given by:

$$S = (\text{mV ml mg}^{-1})$$

The second group of detectors includes those in which the quantity measured is independent of sample concentration in the carrier gas, and is a function solely of the mass flow rate. The sensitivity of these mass flow rate detectors is therefore defined in terms of the following units:

$$S = (\text{A s g}^{-1}) = (\text{Coul g}^{-1})$$

These sensitivities, which may be calculated from the peak area, the sensitivity of the recorder, the reciprocal of the chart speed, the carrier gas flow rate and amount of sample, or in the case of mass flow rate detectors, from the recorder sensitivity, the reciprocal of the chart speed and the peak area, are of no significance in the practical aspects of trace analysis if the noise level of the complete gas chromatographic unit has not been taken into account.

1.313. Noise level, detectability and lower detectable limit

In gas chromatographic trace analysis, as in all high sensitivity methods of measurement where the sensitivity of the detector is utilized to its limit, the recorded signal rarely shows a stable base line. In the case of gas chromatography, fluctuations of the base line are caused both by the electrical components of the instrument, e.g. the recorder and the amplifier (noise) and by mechanical effects, e.g. column bleeding, fluctuations in pressure and flow rate of the carrier gas and the gas supply to the FID, as well as temperature effects (drift). The overall result is generally expressed in terms of the noise level (N). The lower this noise level (dimension: mV) the smaller is the amount of a trace component that can be measured for a given S-value: this quantity is called the detectability (Δ).

$$\Delta = \frac{N}{S} (\text{mg ml}^{-1}) \qquad \text{or} \qquad (\text{mg s}^{-1}) \tag{1}$$

For practical trace analysis the expression given in Equation 1 is not sufficient since a trace component can only be recognized as such, and estimated, when the magnitude of the resulting signal reaches twice the amplitude of the noise level. It is therefore considered that the only decisive information is given by the *lower detectable limit* (Δ_{min}).

$$\Delta_{min} = \frac{2N}{S} \, (\mathrm{mg\ ml^{-1}}) \qquad \mathrm{or} \qquad (\mathrm{mg\ s^{-1}}) \qquad (2)$$

Nevertheless, exceptions are admissible if the pattern of the noise differs significantly from that of the signal. The following diagrams (Figs. 1 and 2) represent two extreme cases of this.

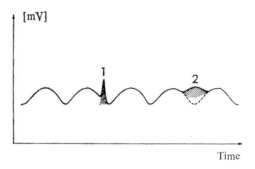

Fig. 1   Determination of trace components in the presence of low frequency noise.

It is seen from Fig. 1, representing low frequency noise, that as a result of its pointed shape the peak for trace component 1 is easily distinguished from the noise pattern and can be measured, although the requirements of Equation 2 are not satisfied. However, component 2 does satisfy this equation,

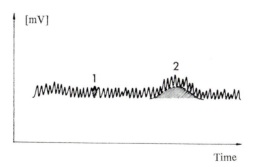

Fig. 2   Determination of trace components in the presence of high frequency noise.

but because it has the same shape and width as the noise signals the peak cannot be identified as a trace component.

Figure 2 illustrates the reverse case where, as a result of the different noise pattern, peak 2 can be evaluated but not peak 1.

TABLE 1 The lower detectable limits for some detectors
(from C. H. Hartmann, *Anal. Chem.*, **43** (2), 113A (1971))

Type of detector	S	N	$2N/S$
Thermal conductivity 250 mA: 220°C 20 ml min^{-1} He	$S = \dfrac{\text{peak area (mV min) . carrier gas flow (ml min}^{-1})}{\text{sample weight (mg)}}$ $= 10\,000$ (mV ml mg^{-1})	0·01 (mV)	$2 . 10^{-6}$ (mg ml^{-1})
Flame ionization 300 V: 20 ml min^{-1} H$_2$ 300 ml min^{-1} air	$S = \dfrac{\text{peak area (A sec)}}{\text{sample weight (g)}}$ $= 0·01$ (Coul g^{-1}) (for SE 30 stationary phase at 200°C)	10^{-14} A	$2 . 10^{-12}$ (g sec^{-1})
Electron capture 35 ml min^{-1} N$_2$ Aldrin	$S = \dfrac{\text{peak area (A min) . carrier gas flow (ml min}^{-1})}{\text{sample weight (g)}}$ $= 800$ (A ml g^{-1}) for a ^{3}H source $= 40$ (A ml g^{-1}) for a ^{63}Ni source	$8 . 10^{-12}$ A $2 . 10^{-12}$ A	$2 . 10^{-14}$ g ml^{-1} 10^{-13} g ml^{-1}

Detector	Sensitivity	Conditions		
Helium	$S = \dfrac{\text{peak area (A sec)}}{\text{sample weight (g)}}$ $= 100$ (Coul g^{-1})	60 ml min^{-1} He 4000 V cm^{-1} Methane	$2 . 10^{-12}$ A	$4 . 10^{-14}$ (g sec^{-1})
Alkali flame ionization	$S = \dfrac{\text{peak area (A sec)}}{\text{sample weight (g)}}$ $= 20$ (Coul g^{-1})	37 ml min^{-1} H$_2$ 220 ml min^{-1} air Parathion, P	$3 . 10^{-14}$ A	$3 . 10^{-15}$ (g sec^{-1})
Flame photometer	$S = \dfrac{\text{peak area (A sec)}}{\text{sample weight (g)}}$ $= 400$ (Coul g^{-1}) for sulphur $= 470$ (Coul g^{-1}) for phosphorus	80 ml min^{-1} N$_2$ 170 ml min^{-1} H$_2$ 20 ml min^{-1} O$_2$	$4 . 10^{-10}$ A	$2 . 10^{-12}$ g sec^{-1} $1 \cdot 7 . 10^{-12}$ g sec^{-1}

Thus exceptions to Equation 2 are admissible if:

$$\frac{\text{peak width of signal}}{\text{peak width of noise}} \ll 1$$

(for low frequency noise)
or

$$\frac{\text{peak width of signal}}{\text{peak width of noise}} \gg 1$$

(for high frequency noise).

The lower detectable limits for trace analyses using the most important detectors are compared in Table 1.

References
1. LOVELOCK, J. E., *Anal. Chem.*, **33**, 162 (1961).
2. GOGH, T. A. and WALKER, E. A., *Analyst*, **95**, 1 (1970).
3. HALÁSZ, I., *Anal. Chem.*, **36**, 1428 (1964).
4. HARTMANN, C. H., *Anal. Chem.*, **43** (2), 113A (1971).
5. DIMBAT, M., PORTER, P. E. and STROSS, F. H., *Anal. Chem.*, **28**, 290 (1956).

1.32. Recorder Presentation of Data

Electronic integrators or on-line data handling systems connected directly to the gas chromatographic detector (thermal conductivity cell, flame ionization detector, etc.) provide less distorted signals than recorders, since the former have a considerably shorter response time and do not exhibit some of the phenomena, such as overshoot, which occasionally occur with recorders. Moreover, manual interpretation of chromatograms, particularly with narrow peaks, gives rise to errors, which should not be underrated. Nevertheless, the use of integrators in trace analysis gives rise to problems. Depending on the noise level, those which do not incorporate baseline correction can produce unresolvable numerical data from which the concentration of the trace component cannot be deduced. Integrators incorporating baseline correction suppress not only the noise but also the trace signal, even if this is clearly distinguished from the noise.

From these considerations it can easily be seen that such analyses are best performed with recorders, since only when a chromatogram is at hand is it possible to see the overall pattern of the baseline and to examine the signal pattern in the region being measured. For trace analytical investigations the recorder must be specially chosen. It must have a span of ≤ 1 mV, a high sensitivity, and for special cases a short response time, and be subject to as little interference as possible. Some of these points and their influence on the signal quality will be examined more closely below.[1]

In a potentiometric recorder the voltage U_x being measured opposes a potentiometric voltage U_k. The resulting potential difference $U_k - U_x$ is

converted by a vibrator having a carrier frequency of 50 Hz and after amplification by a factor of about 10^7 is fed to an induction servomotor which
adjusts a resistance until $U_k - U_x$ becomes equal to zero. In practice, a small
residual potential difference often remains, but which can be so small that,
in spite of the high amplification, it can no longer cause movement of the
recorder servomotor. This results in a region of uncertainty called the 'dead
band' which is influenced by the recorder sensitivity (response), source
impedance of the signal being measured, damping, and by external electrical
interference. Different kinds of interference can give rise to stepped or
flattened signals which cause considerable distortion, particularly for very
small peaks. Such interference can often simulate stable baselines which the
detector unit has not actually produced.

Stepped signals are mainly caused by low recorder sensitivity. As already
mentioned, in gas chromatographic trace analysis there is only a small
difference between the amplitude of the voltage to be measured and the noise
level of the instrument. This is of the order of μV. If these small voltage
measurements are to be correctly displayed, the recording instrument must
have a highly reproducible response. For potentiometric recorders working
on the Poggendorf principle, response is determined by the recorder span,
the resolution of the measuring potentiometer and the sensitivity of the
servo-amplifier (null amplifier). If we consider a potentiometric recorder
having a span of 0 . . . 1 mV, a chart width of 250 mm and a potentiometer
having 1000 turns, then there is a potential difference of 1 μV across each
turn to oppose the voltage being measured. This corresponds to a recorder
displacement of 0·25 mm. The servo-amplifier must therefore have a sensitivity
of ≤ 1 μV to produce a smooth trace of the peak. An amplifier having a
sensitivity of 2·5 μV would not be suitable under these circumstances since
this would give a resolution of only 2·5 μV $= 0·75$ mm. The peak would then
no longer be recorded as a smooth trace; steps would be formed which would
cause very noticeable distortion, particularly for wide trace peaks as
illustrated in the following diagrams (Figs. 3 and 4).

Consequently for this kind of analysis the recorder must have a sensitivity
of at least ≤ 1 μV. This can only be obtained with a high gain amplifier which
can sense and utilize the resolution of the measuring potentiometer.

Trace analysis makes great demands on both the electronic and mechanical
components of the recorder. If, for example, the chart speed is irregular or
there is backlash in the recorder pen drive mechanism then stepped peaks
will result even if the best electronics are used. The thickness of the line
drawn by the pen must also be mentioned since this can significantly affect
the quality of the trace for very small recorder deflections, assuming that it is
functioning correctly with respect to the points referred to above.

Whilst it is mainly the sensitivity which is responsible for producing a
stepped trace, flattened peaks are generally caused by spurious external
signals.

To achieve the high sensitivity required to detect voltages of 1 μV the servo-amplifier must be designed with a high amplification factor. In practice, such amplifiers are very sensitive to spurious signals. These come predominantly from the 50 Hz mains supply and reach the amplifier input through capacitive pick-up or *via* the signal itself. Spurious signals can have values up to a few volts and can therefore overload the amplifier. (Even a few μV a.c. can produce considerable interference on a 1 mV recorder!)

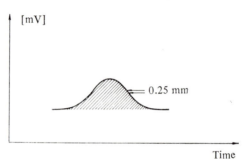

Fig. 3   Smooth display of a broad trace peak.

Since these spurious signals and the sample peak occur simultaneously, the latter cannot be completely resolved. This can result in the reproducibility dropping from 0·1 % to as much as 20 % of the recorder span, so that small peaks are not registered and larger peaks are truncated (*see* Figs. 5 and 6).

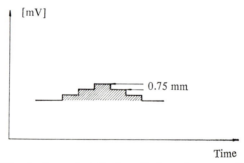

Fig. 4   Imperfect display of a broad trace peak.

In the newer potentiometric recorders, spurious a.c. signals of this kind are suppressed by up to five times the span with ripple filters arranged in series, and which eliminate the interference occurring in practice.

In trace analysis, the response time of the recorder can also be important if rapidly eluted trace components are involved. The response time of a potentiometric recorder is defined as the time which the servo-system needs to move the pen carriage from 0 to 100 % (250 mm) of the recorder span.

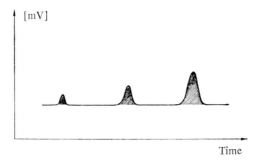

Fig. 5   Undistorted display of trace components.

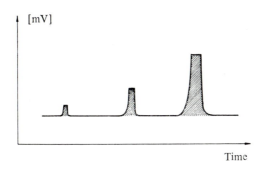

Fig. 6   Imperfect display of trace peaks caused by spurious signals.

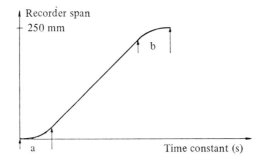

Fig. 7   Variation of the pen carriage speed across the full span of a potentiometric recorder.

In gas chromatography, the source impedance of the particular detector system employed (thermal conductivity cell, flame ionization detector, etc.) must be taken into account since it affects the response time. The speed of the pen carriage across the recorder span is not linear since allowance must be made for acceleration and deceleration periods. Figure 7 shows that in both of these critical regions, *a* and *b*, the recorded signal is too small, so that the results obtained are too low.

Errors will occur if the source impedance and output voltage of the detector unit are not matched with the input impedance of the recorder. The reason is as follows:

In the balanced state, potentiometric recorders have a significantly higher input impedance (~ 2 $M\Omega$) than in the unbalanced state (~ 2 $K\Omega$) and therefore load the source. If a signal being measured has too high a source impedance then the signal is loaded by the potentiometric recorder, and it will assume a lower value.

Depending on the time constant of the recorder, part of the signal being measured is counterbalanced and the loading will therefore become smaller. The input signal then increases with the source impedance. The recorder again counterbalances the new value to an extent which depends on the time constant. This process is repeated until complete compensation of the input signal is attained. Naturally the adjustments occur in small steps which in turn depend on the sensitivity of the recorder. The time taken to reach a state of equilibrium becomes significantly greater if the matching is incorrect, since the time constant of the servo-system is greater for this multi-step compensation. However, small time constants (response times) may be achieved by optimizing the design of the servo-system. Thus, a tacho-generator coupled to the servomotor produces a voltage proportional to the speed, and this is used to control the servo-system.

Reference
1. HACHENBERG, H., *Glass Instrum. Tech.*, 581 (1968).

1.33. Sample Collection, Measurement and Calibration

1.331. Gases

Sample collection from gaseous systems is normally carried out with gas pipettes through which the sample to be analysed is flushed. Rubber or plastic connections must be avoided completely, and that to the sampling point or sample vessel should employ the shortest possible metal-to-glass or glass-to-glass joint. Suitable lubricants are available for glass stopcocks so that the usual effects of grease can be avoided.

Measurement of gaseous samples by compressing them into a known volume using sodium chloride solution or other liquids cannot be used in trace analysis. If necessary, mercury can be used for this purpose. The best method of measurement is by means of the partial pressure at a constant volume.[1] The size of sample volume is determined by the sensitivity of the detector being used, and the efficiency of gas chromatographic separation. Since sensitive detectors are available nowadays it is generally no longer necessary to use significantly larger volumes than for analyses at the percentage level. This has the advantage that separations, often difficult to achieve in trace analysis, can be more easily optimized.

Sample measurement by means of partial pressures is more difficult in trace analysis involving nitrogen, oxygen, argon and carbon dioxide, *i.e.* the components of air (78 vol % nitrogen, 21 vol % oxygen, 0·93 vol % argon and 0·03 vol % carbon dioxide), since the type of vacuum pump normally used does not reduce pressures much lower than 0·5–1 mm of mercury. This means that *e.g.* at 0·7 mm of mercury a blank analysis cannot be less than 780 vol ppm nitrogen, 210 vol ppm oxygen, 9 vol ppm argon and 0·3 vol ppm carbon dioxide. This difficulty can be minimized by careful repeated flushing of the gas measuring unit with the sample, but it cannot be completely eliminated, so that the results of analyses of traces of these constituents using the above method are often inaccurate.

When the gas to be analysed is above atmospheric pressure it is recommended that the sampling be from a suitably filled pressure vessel. However, the usual flushing technique through a calibrated loop in the gas sampling system cannot be employed since, at the ppm level, back diffusion of the constituents of air occurs. One possible method of avoiding this problem is to fit a capillary tube at the outlet of the gas sampling system and immerse the end of it in a suitable liquid, *e.g.* glycol (*see* Fig. 8). The exclusion of air achieved in this way prevents back diffusion.

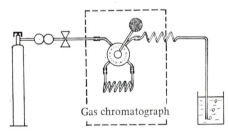

Fig. 8   Measurement of gaseous samples in the absence of air.

The same problems arise with sample collection if analysis involves determination of traces of the gases mentioned above. When the sample pressure is greater than atmospheric, calibrated stainless steel vessels having a valve at each end can be used (*see* Fig. 9).

To collect a sample of the gas being analysed, the pressure vessel is connected to the sampling point with an Ermeto screw coupling and thoroughly flushed with the gas for about half an hour. The exit valve is then closed and the vessel filled by means of the excess pressure in the sampling line. If the gas to be analysed is not above atmospheric pressure then sample collection and measurement can only be carried out by conventional methods using mercury, and this must be absolutely air-free.

With highly polar trace components such as water, methanol, formaldehyde, hydrogen sulphide, sulphur dioxide, ammonia, etc., it is essential to use a discontinuous method of sample collection owing to the strong adsorption effects shown by the walls of the sample container. It is probably advisable

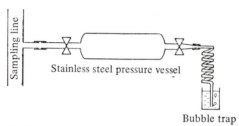

Fig. 9   Collection of gaseous samples in the absence of air.

to use a gas reservoir made of polished aluminium. Analysis of these materials should be carried out by an 'on-line' technique, *i.e.* by connecting the gas chromatograph directly to the sampling line.

Special precautions should be taken in the collection and measurement of samples of liquefied gases. A mistake which is frequently made is to vaporize the sample directly from the sampling line or the sample container into a gas pipette. Depending on the sample composition this can lead to considerable fractionation. It is therefore essential that the sample should be taken in the liquid state. Metallic sample containers having a volume of 2–3 ml can be used for this purpose. These are easily made from a short stainless steel or copper tube (6 mm diameter) and two suitable valves fitted with screw couplings (*see* Fig. 10).

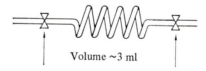

Fig. 10   Device for collecting samples of liquefied gases.

This apparatus may be used in two ways to collect samples of liquefied gas:

(*a*) with one end connected to the sampling point. The sample is collected by bleeding the liquefied gas into the atmosphere. This should be continued until, as a result of self-cooling of the sample coil, only liquefied gas escapes.

Both valves are then closed. This method of sample collection is preferred for liquefied gases taken from pipelines (Fig. 11).

(b) with both ends connected to the source of the sample so that the sampling device constitutes a by-pass (Fig. 12).

Fig. 11   Collection of a sample of liquefied gas.

This method of sample collection is basically more accurate. It should always be used when the contents of a liquefied gas container are to be analysed for trace materials. Before collecting the sample, careful attention must be paid to obtaining uniform material by pumping it through the coil (*see* Fig. 12).

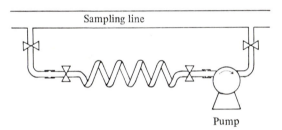

Fig. 12   Collection of a sample of liquefied gas.

The analysis itself is carried out after this liquid sample has been quantitatively vaporized into a small glass gas burette filled with mercury. With chemically reactive trace components, the mercury surface should be covered with a thin layer of a suitable liquid (*see* Fig. 13).

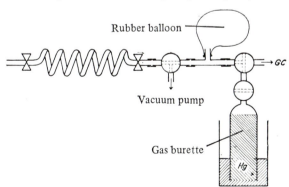

Fig. 13   Vaporization apparatus for liquefied gases.

Calibration with gaseous trace components requires considerable skill. A basic mistake often made which also applies to liquid measurement is to base the calculation of the concentration in ppm or ppb on a calibration performed at the percentage level. This error can be appreciated from the fact that the calibration curve E (Fig. 14) is no longer a straight line in the lower concentration region. This can be caused by both a detector having a non-linear response and adsorption and desorption phenomena which are only apparent at very low concentrations, particularly with polar substances. Consequently, both positive (I) and negative deviations (II) can occur. A positive deviation results when a portion of the trace component adsorbed during the preceding analysis is eluted in the subsequent calibration. This effect is easily checked by means of a blank analysis. A negative deviation is caused by partial adsorption of the trace constituent being analysed.

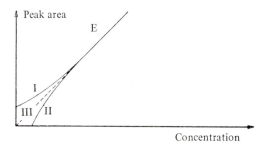

Fig. 14   Non-linear calibration plots in trace analysis.

For the above reasons it is important that a calculated or graphical extrapolation (dashed line III) should not be made from the higher levels to the region of trace concentrations. If this were done then, in the low concentration region, calculations based on the calibration graph III would lead to high results in the case of I and to low results in the case of II, and they could be in error by several hundred per cent.

There is therefore a basic rule in trace analysis that calibration should be performed in approximately the same concentration range as that in which the trace components are actually present. This is especially important when trace components elute in the tail of the main peak. Compared with calibrations at higher concentrations, the procedure is often considerably more time consuming. The best arrangement for routine purposes is to have a large stock of standard mixtures available. Preparation of such mixtures, either with or without mixing pumps, is relatively slow and requires large amounts of pure gases which are often not available. Moreover, when such mixtures are stored for long periods changes in composition may occur. This happens particularly with chemically reactive gases when they are kept in steel pressure vessels. It is therefore advisable to prepare the calibration mixture immediately before use, and if possible to introduce it directly into the gas chromatograph.

Exponential dilution is an example of such a method (Lovelock 1961). In this procedure, a stream of carrier gas flows continuously through a vessel of known volume V. A measured amount of the gas being calibrated is introduced into this vessel with a gas-tight syringe, the vessel being equipped with a stirrer. If the flow rate of the carrier gas is Q then at time t the concentration C_p of the gas being calibrated is given by:

$$C_p = C_o \exp\left(-\frac{Qt}{V}\right) \tag{3}$$

where C_o is the initial concentration at time t_o.

The concentration C_p at a certain time t may therefore be expressed by:

$$\log C_p = \log C_o - \frac{Qt}{2 \cdot 3 V} \tag{4}$$

or the time t may be calculated after which a definite concentration C_p is obtained:

$$t = \frac{2 \cdot 3 V}{Q}(\log C_o - \log C_p) \tag{5}$$

Figure 15 shows schematically the sequence of peaks as they are normally obtained from an exponential dilution cell (Fig. 16).

This method is particularly suited to calibrations over wide ranges of concentration and is extremely simple to use. It enables the linearity of the detector response and the limits of detection to be established simultaneously.

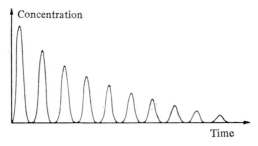

Fig. 15   Diagrammatic representation of the sequence of peaks from an exponential dilution cell.

Unfortunately the technique cannot be used if the calibration of the trace component must be carried out at the same time as that of the main component. This situation arises when the trace material elutes immediately before the main component peak or in its tail. In such cases, calibration may be made by introducing a mixture of the trace material and the main component immediately before the column inlet. The two components are contained in vessels of widely different volumes connected to the gas chromatograph, and are therefore introduced in proportion to their partial pressures. They are subsequently flushed together onto the column by the carrier gas (Fig. 17).[2]

When calibrating very small amounts of materials by this method it is often necessary to first dilute with the carrier gas before introducing the

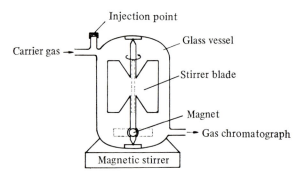

Fig. 16   Schematic diagram of an exponential dilution cell.

material into volume *B*. This is performed in a 1 l or 3 l gas mixing pipette of the type shown in Fig. 18. The pipette is first evacuated, and the trace material being calibrated introduced with a gas-tight syringe. The pipette is then filled with carrier gas and shaken so that the movable glass piston (*a*) produces thorough mixing.

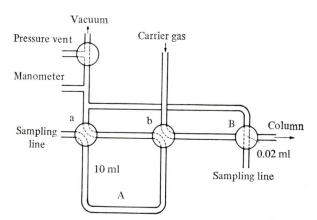

Fig. 17   Sampling system for preparing ppm standard gas mixtures.

The concentration of trace components in this gas mixture may be determined relatively accurately by gas chromatographic analysis, the error usually being 2–5%. Partial pressure introduction of this mixture into *B* (Fig. 17) enables very small amounts of the standard mixture to be measured, and together with the known amount of the main component in A (Fig. 17) it can be injected on to the column by rotating the 4-way stopcocks a and b.

This may be explained in detail by way of a practical example, the preparation of samples having 2 and 20 vol ppm of acetylene in ethylene:

Volume of the large sampling loop A: 10·0 ml.
Volume of the sampling stopcock B: 0·02 ml.

Let the concentration of acetylene in pure nitrogen in the gas mixing pipette be 1 vol %.

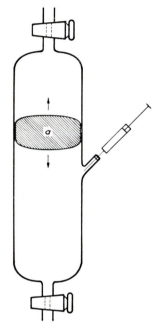

Fig. 18   Pipette for the preparation of gaseous mixtures.

The introduction of this mixture at a pressure of 760 mm of mercury into the 0·02 ml volume gives a measured sample of 0·0002 ml of C_2H_2 or if a pressure of 76 mm of mercury is chosen then the amount of C_2H_2 is 0·000 02 ml. When C_2H_2–free ethylene contained in the 10 ml sample volume is injected onto the column together with one of the above mixtures, then the concentrations based on the 10 ml volume are:

$$20 \text{ ppm } C_2H_2 \text{ in } C_2H_4$$

or

$$2 \text{ ppm } C_2H_2 \text{ in } C_2H_4$$

This calculation is not quite correct since a total gas volume of 10 + 0·02 ml = 10·02 ml is used, whereas in subsequent analyses 10·00 ml of C_2H_4

would be measured out. However, this corresponds to a difference of only 0·2% and may be neglected since the relative error of analysis is greater by two orders of magnitude. The precision of this sampling method is demonstrated for very small sample volumes in Fig. 19.

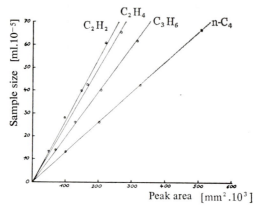

Fig. 19   Dependence of the peak area on the sample size for acetylene, ethylene, propylene and n-butane in the range $0·5 \times 10^{-5}$ to 70×10^{-5} ml.

Another similar method of preparing ppm concentrations of compounds uses the glass apparatus shown in Fig. 20. This consists of a normal gas mixing pipette fitted with a micro-volume glass stopcock.[3] With this apparatus the required amount of a pure trace component or an appropriately diluted

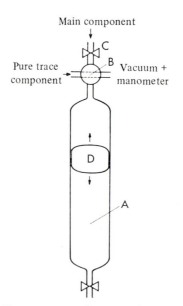

Fig. 20   Glass apparatus for preparing ppm mixtures.

mixture in a suitable diluent can be measured out, as already described, by introducing a known pressure of gas into the small volume *B*. The large volume *A* of the mixing pipette is first evacuated. By turning the glass stopcocks *C* and *B* both the trace and main components are introduced into *A* where they are mixed by means of the glass piston *D*.

The methods described so far are of only limited use for highly polar and chemically reactive gases and vapours owing to adsorption on the glass walls.

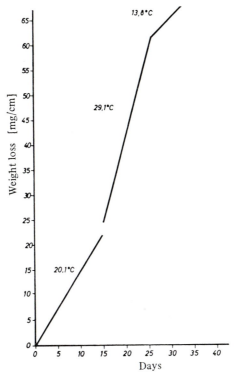

Fig. 21   Permeation of NO$_2$ through FEP Teflon (from A. E. O'Keeffe and G. C. Ortman, *Anal. Chem.*, **38**, 763 (1966)).

A universally applicable method of calibration is based on the permeability of certain materials to gases and vapours. O'Keeffe and Ortman use a short Teflon tube, closed at each end with a metal or glass ball, to contain the gas being calibrated in the condensed state.[4] Permeation of the gas is controlled only by the temperature and the material from which the tube is made. The rate of permeation may be determined by measuring the loss in weight of the tube over a definite time interval. Figure 21 shows this loss in weight over a period of 40 days for NO$_2$ from an FEP Teflon tube at three different temperatures.

For the preparation of standard mixtures, the tube can be immersed in an accurately known volume of gas or a stream of gas can be allowed to flow around the Teflon tube at fixed temperatures. This method enables precisely known concentrations, *e.g.* of propane in air, to be obtained by variation of the flow rate. During preparation of the mixture, the propane concentration is continually measured with a flame ionization detector. In this kind of calibration, attention must be paid to the possibility of an erroneous permeation rate owing to peculiarities of the tube wall. The humidity of the surrounding gas must be kept very constant, since sorption and desorption of water on the wall of the tube can cause fluctuations in the permeation rate. In tests with NO_2 microscopically small blisters form on the wall of the tube

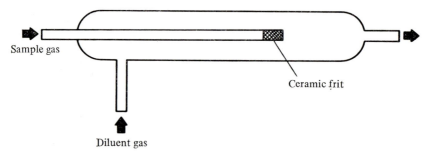

Fig. 22   Device for preparing ppm concentrations using a ceramic frit (from H. D Axelrod, R. J. Teck, J. P. Lodge Jr and R. H. Allen, *Anal. Chem.*, **43**, 496 (1971)).

and these also influence the permeability. In the case of hydrogen sulphide, oxygen which has diffused inwards can result in the formation of colloidal sulphur in the wall of the tube, and this produces the same effect. In addition, some substances such as SF_6 and CF_3Br cause swelling of the wall of the tube. For the preparation of ppm standard mixtures Axelrod *et al.* have used a ceramic frit which was connected to a cylinder containing the gas under pressure.[5] In this method, the diluent gas flows around the frit (Fig. 22). This permits any desired concentration to be prepared by variation of the pressure before the frit and of the flow rate of the diluent gas.

Another method which has been described depends on the rate of diffusion of vapour through a glass capillary tube into the gas stream.[6] A diffusion cell based on this principle is shown in Fig. 23. The lower approximately 50 ml capacity round-bottom flask contains the liquid, the vapour of which is used to prepare the calibration mixture. The space above the liquid becomes saturated with the vapour and it diffuses through the capillary tube into the similar-sized upper flask, in which it mixes with a stream of diluent gas, and is then discharged.

For a given apparatus and a particular liquid, the rate of diffusion depends only on the total gas pressure and the vapour pressure of the liquid. If the total pressure is constant, then the rate of diffusion is a function only of temperature and can therefore be controlled by surrounding the cell with a thermostat bath. Although for gases and vapours having known diffusion coefficients the rate of diffusion can be calculated from the dimensions of the cell, for accurate work it is better to calibrate the apparatus. For this purpose the loss in weight of the lower flask, after passing the diluent gas for a certain time, is measured at a precisely controlled temperature.

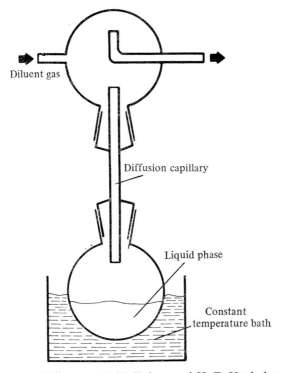

Fig. 23   Diffusion cell (from J. M. McKelvey and H. E. Hoelscher, *Anal. Chem.*, **29,** 123 (1957)).

Calculation of the rate of diffusion is based on the following equations which can also be used to deduce optimum values for the diameter and length of the capillary tube and temperature, for particular applications:

$$r = \frac{DPMA}{RTL} \ln \pi \qquad (6)$$

where

r rate of diffusion
M molecular weight of the vapour
P total pressure
A cross-sectional area of the capillary tube

L length of the capillary tube
T absolute temperature
R gas constant
D diffusion coefficient
π $(P/P - p)$
p vapour pressure of the liquid

The values at other temperatures and pressures may be calculated from

$$D = D_0\left(\frac{T}{T_0}\right)^2 \left(\frac{P_0}{P}\right) \tag{7}$$

where

D diffusion coefficient at temperature T and pressure P
D_0 diffusion coefficient at 0°C and 1 atm.

Finally, combination of these two equations gives

$$r = \left(\frac{D_0 P_0 MA}{RT_0{}^2 L}\right) T \ln \pi \tag{8}$$

The difference between the calculated and experimentally determined values of the rate of diffusion is *e.g.* 5% for a toluene–air mixture (calculated $1{\cdot}21 \times 10^{-6}$ g min^{-1}, experimental $1{\cdot}15 \times 10^{-6}$ g min^{-1}). Altshuller *et al.* discussed the experimental limits of these diffusion cells, and pointed out sources of error which can explain this difference between the calculated and experimental values.[7] Thus, one of the sources of error is turbulence at the end of the capillary tube and the effect is enhanced as the flow rate of the diluent gas is increased. Altshuller therefore recommended the type of diffusion cell shown in Fig. 24 since this design enables accurate measurements to be made.

The advantage of this arrangement is that the diffusion tube itself serves as the reservoir for the liquid. As a result of this, calibration may be based on changes in weight or volume, the tube being graduated for the latter purpose.

For the calibration of vapours of polar liquids, even the methods just described involve errors owing to adsorption. For the calibration of such vapours, Hachenberg has recommended, as already described, separate, but simultaneous sampling of the trace and main components, the former being measured in the liquid rather than the vapour state.[8] The gas chromatograph employed must therefore be equipped with both a gas sampling valve and a liquid sampling system. Preparation of the ppm mixture is carried out at the column inlet by injecting the polar trace component as a liquid at the same time as the gaseous main component is introduced. If required, the polar substance may be mixed with a diluent to which the detector does not respond (*e.g.* water for the FID). The polar compound will vaporize and become part of the total gas volume, *i.e.* it must be added to the volume of the

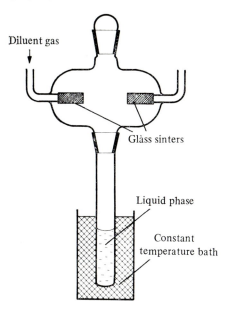

Fig. 24   Diffusion cell (from A. P. Altshuller and I. R. Cohen, *Anal. Chem.*, **32**, 802 (1960)).

diluent gas. This method of calibration does not require special mixing equipment, it is easily performed, and the only error involved is that due to the syringe. Figure 25 illustrates the procedure for calibrating methanol vapour in ethylene.

Calibration of traces of gases in liquids is another rather difficult task which is frequently encountered in petrochemistry. The following method is recommended for this purpose.[9] A gas pipette of known volume is filled with the gas being analysed, and then cooled in a low temperature bath. This produces a reduced pressure in the pipette, and the liquid, in which the gas is to be

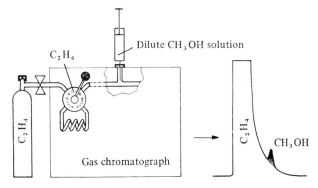

Fig. 25   Method of calibrating organic vapours in gases.

determined, is drawn into the pipette until it is completely filled. The next step is simply to prepare the required ppm mixture by appropriate dilution with the gas-free liquid. Obviously in this method it is assumed that the gaseous and liquid components are completely miscible.

1.332. Liquids

The sampling of liquids generally involves fewer problems than gases. However, trace analysis of water is an exception since it is present universally, and therefore special precautions are necessary. Compared with gases, the preparation of ppm mixtures of liquids is far simpler since they are easier to handle. A difficulty which occasionally arises is that an insufficient amount of the main component is available for preparation of the mixture. This can be overcome by using another suitable liquid to make up the required solution.

Standard ppm mixtures of liquids may be prepared more accurately by volume measurement than by weight, since there is considerable error in weighing the small amounts involved. Special precautions must be taken in the preparation of ppm concentrations of water in liquids, particularly when these standard solutions are to be kept for long periods. Measuring flasks with glass stoppers should be avoided if possible since they allow atmospheric water vapour to leak in. Cotton *et al.* have designed a measuring flask for this purpose which has a Teflon stopper and an internal diameter of 1 mm at the neck.[10]

CALIBRATION

Calibration of traces of liquid substances may be performed by the reproducible injection of the appropriate ppm mixtures with precision syringes or by the internal standard method, in which care must be taken to adjust the concentration of the standard substance to match the order of magnitude of the concentration of the trace component being calibrated.[11] For polar compounds this procedure, which at first sight appears to be very simple, is in fact complicated by 'memory effects'. These can be caused by difficult to control irreversible adsorption phenomena occurring in the sampling system (rubber septa), or on the walls of the chromatographic column, the solid support, or the stationary phase. Rubber seals in tube connections in the gas chromatograph and injection septa are also considered to be major sources of trouble. Depending on the material from which they are made and their quality, septa can dissolve and subsequently liberate appreciable amounts of organic compounds and therefore cause considerable interference, particularly in temperature programmed trace analysis. Callery therefore recommended improving the stability of the septum by covering its surface with PTFE or fibreglass.[12] Rubber septa may also be made more stable by heating at 300°C in a stream of nitrogen for a period of ninety hours.[13]

In addition to the effect of the septum, the design of the sampling system is important. Thus, Dressman showed for phenols and organic acids as model compounds that 'memory' peaks are mainly caused by the dead space in the sampling system and therefore recommended a method by which it is possible to inject directly onto the column.[14] In Fig. 26 an injection system (*A*) which is not very suitable for trace analysis is compared with the design proposed by Dressman. Another part of the gas chromatograph where this problem occurs is the connection between the column exit and the detector.

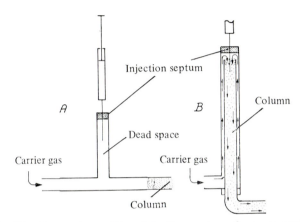

Fig. 26   Comparison of two injection systems for trace analysis.

(A) unsuitable   (B) suitable (from R. C. Dressman, *J. Chromatog. Sci.* **8**, 265 (1970)).

This must be as short as possible, efficiently heated and easily replaceable for cleaning. For all of the above reasons, the whole of the gas chromatographic equipment should be tested for contamination before each trace analysis or calibration. This is best achieved by means of a 'blank' analysis carried out with an FID by injecting a polar liquid such as water, formic acid or a mixture of these two substances in the ratio 1:1. Only if no peaks are produced is the instrument ready for use. It is thus appropriate to conclude that gas chromatographs which are normally employed for all kinds of analyses can be used for trace analysis when the occasion arises.

References
1. HORN, O., SCHWENK, U. and HACHENBERG, H., *Brennstoff-Chem.*, **39**, 336 (1958).
2. SCHWENK, U., HACHENBERG, H. and FÖRDERREUTHER, M., *Brennstoff-Chem.*, **42**, 194 (1961).
3. BERINGER, K., unpublished work.
4. O'KEEFFE, A. E. and ORTMAN, G. C., *Anal. Chem.*, **38**, 760 (1966).
5. AXELROD, H. D., TECK, R. J., LODGE, J. P. Jr and ALLEN, R. H., *Anal. Chem.*, **43**, 496 (1971).

6. McKELVEY, J. M. and HOELSCHER, H. E., *Anal. Chem.*, **29**, 123 (1957).
7. ALTSHULLER, A. P. and COHEN, I. R., *Anal. Chem.*, **32**, 802 (1960).
8. HACHENBERG, H., unpublished work.
9. GUTBERLET, J., unpublished work.
10. COTTON, T. M., BALLSCHMITTER, K. and KATZ, J. J., *J. Chromatog. Sci.*, **8**, 546 (1970).
11. HAMILTON, C. H., in *Instrumentation in gas chromatography* (Ed. J. Krugers), Centrex Publishing Co., Eindhoven (1968), Chapter 3.
12. CALLERY, I. M., *J. Chromatog. Sci.*, **8**, 408 (1970).
13. KOLLOFF, R. H., *Anal. Chem.*, **34**, 1840 (1962).
14. DRESSMAN, R. C., *J. Chromatog. Sci.*, **8**, 265 (1970).

1.34. Concentration Techniques

It becomes necessary to concentrate trace components when the detector sensitivity is no longer adequate, or when sufficient separation of trace components from other constituents in the sample cannot be achieved. There are two principal aims for concentration procedures. The first arises if a trace analysis must be carried out where the identity of the trace components is unknown, so that they must first be isolated. Large samples are necessary since enrichment factors from 1:1000 to 1:10 000 must be achieved in order to have sufficient material available for a clear identification by other analytical methods. In general, the following methods of concentration may be considered for this purpose:

Distillation,
Preparative gas chromatography,
Removal of the main component by chemical reaction, and in some cases by condensation and adsorption.

The second aim of enrichment is to facilitate the accurate quantitative determination of known trace components, such as may be necessary in routine analyses. Methods of concentration in this case are naturally different since they must be carried out more rapidly and most require enrichment factors of only from 1:10 to 1:1000. The following methods should be mentioned:

Separation by gradient elution chromatography,
The technique of peak cutting.

Since some of these methods can only be used in special cases, it is often necessary to use combinations of the concentration techniques mentioned above to achieve the objective.

1.341. Distillation and preparative gas chromatography

Enrichment for purposes of identification by condensation or adsorption of trace components at the exit of the analytical column can only be carried out in exceptional cases since the sample size and consequently the amount of the trace component available is severely limited.[1] Efficient distillation or preparative gas chromatography is therefore employed. The latter is superior to almost all other methods of separation such as spinning-band distillation of liquids and low temperature spinning-band or Podbielniak distillation of gases for the reasons given below.

Preparative gas chromatography can be used to separate azeotropic mixtures.

Gas chromatographic separation takes place considerably faster and can be performed under milder conditions than distillation. In the latter case, for example, dissociative loss of HCl can occur with chlorinated compounds, and with highly efficient distillation columns the mixture being separated is kept for long periods at high temperatures in the still pot.

Preparative gas chromatography can also be used to separate those samples which cannot be distilled owing to shortage of material. This difficulty is due to the fact that, with a small amount of material, the hold-up of the distillation column is comparable to the volume of material to be distilled.

A disadvantage of preparative gas chromatography compared with distillation methods is, however, that only a limited amount of material can be processed in one pass. Also it is very difficult, and sometimes impossible at the high carrier gas flow rates of $50-70\,l\,h^{-1}$ to isolate from a multi-component mixture an unknown compound present at a concentration of $\leq 1\%$. The vapour pressure of this small amount is reduced to such an extent that the quantity usually obtained is insufficient even for mass spectrometry. In this case, it may be possible to trap the trace component in a solvent, a technique which is often used with analytical instruments (carrier gas flow rate $3-7\,l\,h^{-1}$) to isolate the major components. If carbon tetrachloride is used, the identification may then be carried out by infrared spectroscopy.

However, with preparative apparatus where the carrier gas flow rate is $50-70\,l\,h^{-1}$ this technique is difficult since relatively large volumes of solvent must be used. This means that the concentration of the trace component in the solvent is frequently too low for subsequent identification. In this case, it may be possible to combine this component with a substance which is not present in the original mixture by choosing an additive having the same retention time as the component in question, and collecting the two compounds together.[2] The trace component may be isolated by this means, but it is in the form of a mixture with the additive. Nevertheless, gas chromatographic qualitative analysis of this two-component mixture on several columns is considerably easier than that of the original multi-component mixture. Moreover, for a two-component mixture mass spectrometric analysis is relatively simple, and in favourable cases even infrared analysis is possible. The following example is given as a simple illustration of this method:

To a mixture of 56·9 wt % acetone, 42·1 wt % toluene and 1 wt % benzene was added 10% diethyl ketone which has the same retention time as benzene

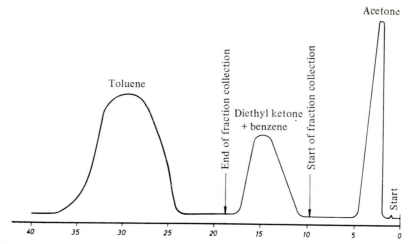

Fig. 27   Chromatogram of a mixture of acetone, benzene, diethyl ketone and
toluene on a preparative 3 m squalane column.

on the squalane column employed. Benzene and diethyl ketone were separated
from this mixture by gas chromatography. The quantity collected was
sufficient for the examination of this two-component mixture on analytical
columns. Figure 27 shows the chromatogram of the preparative separation
on a squalane column, and Fig. 28 shows the two-component mixture
resolved on a sebacic acid ester analytical column.

This technique is not always successful, and consequently preparative
gas chromatography is usually only used for pre-concentration, while actual

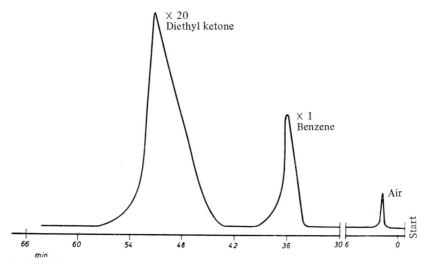

Fig. 28   Chromatogram of a mixture of benzene and diethyl ketone on a 3 m sebacic
acid ester column.

isolation of the trace component is performed on analytical equipment. For gases, this pre-concentration may be carried out using the simple Janák method with a nitrometer as detector and CO_2 as carrier gas.[3,4] Compared with preparative gas chromatography using flame ionization or thermal conductivity detectors, the Janák method clearly has the following advantages:

The apparatus is easily constructed from simple components, and is relatively insensitive to fluctuations in flow rate and temperature. Bleeding of the stationary phase does not interfere since the eluate is not being collected in a cold trap, but over an aqueous KOH solution which will absorb any of the stationary phase, and also water, carried along with the components. In these circumstances, stationary phases such as acetonyl acetone, which are highly selective for gases, may be used. The inconvenient condensation of gases in cold traps is completely avoided since the gas concerned may simply be transferred from the nitrometer into a gas cell or gas pipette for further analysis. In contrast with the condensation method, even low percentage concentrations of the components being examined may be approximately quantitatively determined, since the carrier gas is absorbed by the KOH solution.

A disadvantage of Janák's method is its restriction to gases which do not react with KOH. In order to be able to work on a preparative scale, it is necessary to increase the size of the analytical apparatus by a factor of ten, and to make some other modifications.[5] Thus, for example, the nitrometer will then have a volume of about 5 l and must be provided with a cooling coil because of the considerable heat of neutralization formed in the reaction of $60–70 \, l \, h^{-1}$ CO_2 with 45% aqueous potassium hydroxide solution. However, this coil also improves the dispersion of the bubbles of CO_2 entering the nitrometer. Figure 29 shows the apparatus which, in contrast to a normal analytical nitrometer, is fitted with a rotatable distribution manifold having four burettes.

With this apparatus, the main components may be conveniently removed and the trace components isolated, and also fractions of multi-component mixtures concentrated. A further use of the equipment is for the removal of the main component by chemical reaction. Thus, for example, in the case of ethylene containing traces of N_2O, inert gases and other inorganic constituents can be concentrated if concentrated sulphuric acid saturated with silver sulphate, or bromine is used to fill the nitrometer. Figure 30 shows the gas chromatographic analysis of a fraction concentrated from ethylene in which N_2O having an initial concentration of 0·5 vol ppm has been enriched to 3000 vol ppm.[6]

1.342. Concentration in the chromatographic column (chromatographic sorption)

The simplest method of concentration is to use a short column packed with an adsorbent or a stationary phase on an inert solid support. The column can

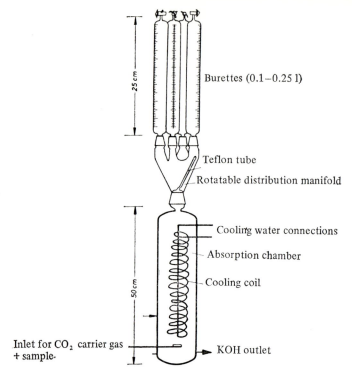

Fig. 29   Diagrammatic representation of a 5 l nitrometer.

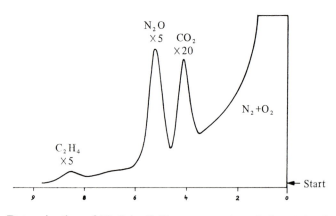

Fig. 30   Determination of N_2O in C_2H_4 on an activated charcoal column after concentrating over Ag_2SO_4/H_2SO_4 solution.

be U-shaped or a spiral and constructed so that it may be cooled in a Dewar vessel (Fig. 31). These traps are mainly used when trace components have to be concentrated away from the laboratory. Elution of the components collected is subsequently carried out in the laboratory by heating the tube while it is being flushed with gas so as to introduce the sample into the gas chromatograph. This method has the drawback that there is no, or only very limited, preliminary separation during the enrichment step since the precolumn is not connected to a detector. Moreover, the scope of the technique is limited by the unfavourable ratio of the volatility of a compound to its partition coefficient, which decreases as the volatility increases. As a

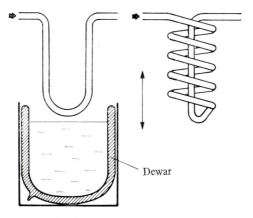

Fig. 31   Condensation devices.

consequence, the maximum volume of gas which can possibly be used without any loss of material is controlled by the most volatile component in the sample. Janák therefore recommended that equilibrium should be established between the sample and the stationary phase in the precolumn.[7,8] By appropriate choice of partition medium in the precolumn, it is possible to achieve selective concentration of certain components in a mixture, and to prevent the influx of undesirable or interfering constituents such as water. The physical basis of this technique is as follows:

During sample collection, a chromatographic frontal process takes place in the equilibrium tube. Equilibrium between the phases is established when the sample gas has the same composition at the tube outlet and inlet. From the amount of a component absorbed and of stationary phase in the tube, it is possible to deduce the concentration of the trace component being analysed if its partition coefficient is known. It is of course necessary previously to have determined this coefficient value from the specific retention volume of the compound on the particular stationary phase employed. Alternatively, retention data from the literature may be used.

The concentration of trace component may be calculated from the expression:

$$C_G = \frac{273\, m}{V_g W_L T} \qquad (9)$$

where

m weight of the trace component (g)
W_L weight of the stationary phase (g)
V_g specific retention volume of the trace component for the stationary phase employed (ml g^{-1})
T temperature
C_G concentration in the gas phase

Equilibration between phases occurs in a short glass tube (3–5 cm × 0·5 cm) filled with a precisely weighed amount of an appropriate packing (Fig. 32).

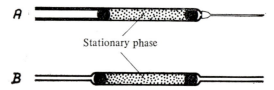

Fig. 32   Equilibration tubes (from J. Janák, *TIPS, 32 GC*, Bodenseewerk, Perkin-Elmer).

Tube *A* is suitable for collecting samples in the laboratory, and tube *B* for use outside the laboratory. When used for this purpose, *B* may be sealed after collecting the sample by melting both ends. Figure 33 shows, as a typical example, the determination of trace components in coal gas.

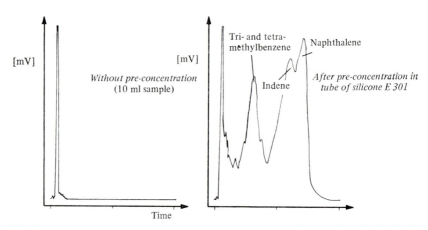

Fig. 33   Chromatogram of trace components in coal gas (from J. Novák, V. Vašák and J. Janák, *Anal. Chem.*, **37,** 660 (1965)).

1.343. Concentration by temperature gradient chromatography

Temperature gradient chromatography is a chromatographic process in which there is a temperature gradient along the stationary phase in the direction of separation, *i.e.* in the direction in which the mobile phase progresses. In contrast to isothermal chromatography, where separation of components results from different solubilities in the stationary phase at constant temperature, separation in chromathermography depends on the solubility in the stationary phase at different temperatures. The chromatogram for chromathermography therefore represents something different from that for isothermal, or even for temperature programmed gas chromatography. The experimental equipment incorporates a tubular heater, having a temperature gradient which is as linear as possible, travelling along the column in the direction in which the mobile phase moves (Fig. 34).

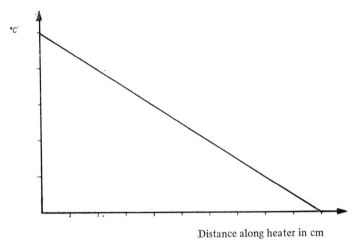

°C

Distance along heater in cm

Fig. 34   Temperature variation along the length of the tubular heater.

One of the first versions was cyclo-chromathermography, in which a concentric heater having a temperature gradient which decreased in the direction of motion, was moved around a column packed with an adsorbent,[9] as shown in Fig. 35.

Kaiser improved and extended this basic principle both by changing the geometrical configuration of the column and by locating auxiliary cooled zones at its inlet and outlet.[10] He termed this system reversion gas chromatography.

The straight, vertical column is surrounded by a movable concentric heater. Compared with the configuration shown in Fig. 35 this is more symmetrical and consequently achieves better resolution. The cold zones at the inlet and outlet of the column prevent the possibility of continuous

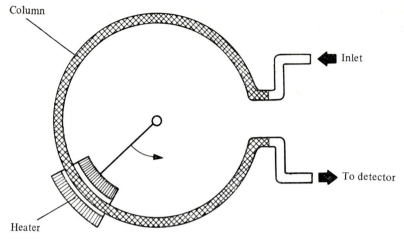

Fig. 35   Schematic representation of cyclo-chromathermography.

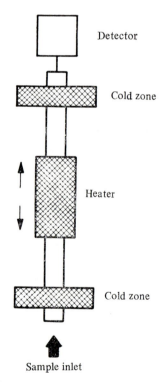

Fig. 36   Operating principle of a reversion gas chromatograph.

escape of low boiling components which would increase the background signal from the detector. This ensures that the maximum sensitivity available from the detector can be utilized. The inlet to the column (Fig. 36), which in this system is called a 'hold-up' column, must be sufficiently cold to quantitatively absorb those compounds which are admitted with the continuously flowing mobile phase, *i.e.* to prevent penetration of the compounds being analysed. The hold-up time can be varied from minutes to hours and determines the total time for the analysis, and the degree of enrichment. At the end of this hold-up time, the heater is used to raise the temperature of the cold zone. When the column inlet has reached the required temperature, the concentric heater, which has a temperature gradient along it, is set in motion. The components being separated advance with the heated zone to the point where, for each particular one, the temperature is just high enough to ensure a sufficiently rapid migration. The rate of migration also depends on the flow rate of the mobile phase and the nature of the stationary phase. Since the different temperature zones reach the end of the column at different times, then the individual compounds similarly arrive at the end of the stationary phase and are carried with the mobile phase into the detector. Depending on the hold-up time, this procedure may be repeated as often as desired, hence the name reversion gas chromatography.

In contrast to the usual gas chromatographic concentration techniques in which enrichment can be achieved by cooling at the inlet end of the column or by means of separate low temperature columns, in temperature gradient gas chromatography the original concentration of substances entering the column is greatly increased in the moving heated zone. Those molecules which precede the peak front quickly enter a region of linearly decreasing temperature, and those components located in the tail soon enter a region of linearly increasing temperature. Consequently, the peaks become very narrow so that a further increase in sensitivity is obtained. Reversion gas chromatography is therefore significantly more sensitive for volatile compounds in gases than any other method of trace analysis so far developed. Kaiser reported an increase in the sensitivity by a factor of 1000 and a limit of detection of about $5 \times 10^{-10}\%$.

Chromathermographic separation may be further improved by coupling the reversion gas chromatograph with an isothermally operating gas chromatograph.[11] However, it should be pointed out that the retention times obtained with this combined unit are not the same as those for the gas chromatograph under isothermal conditions, since a preliminary separation has taken place in the reversion gas chromatograph. Figure 37 shows the basic arrangement for this combination.

A further important advantage of reversion gas chromatography is that it involves continuous sampling, which if possible should always be used in trace analysis as it largely avoids the uncontrollable adsorption phenomena able to occur when the sampling is discontinuous.

The method of continuous sample injection used in reversion gas chromatography depends on the physical state of the sample. In the analysis of gases, they can themselves be used as the carrier gas. Obviously the correct type of detector must be employed. Thus, for example, in the analysis of air, where traces of hydrocarbons have to be detected, a flame ionization detector is the best choice. Liquids are introduced into the column by bubbling an

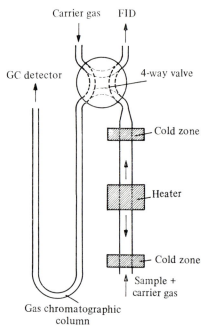

Fig. 37   Basic arrangement for coupling a reversion gas chromatograph with an isothermally operating gas chromatograph.

appropriate carrier gas through them, and they are therefore accompanied by the volatile constituents. A similar procedure is used for the analysis of volatile materials in solids.

1.344. Peak-cutting as a means of increasing the detection sensitivity

Although the sensitivity of currently available detectors enables the quantitative analysis of trace components to be carried out in most cases without using pre-concentration procedures, it repeatedly happens that the maximum sensitivity of the detector cannot be utilized owing to inadequate resolution. This situation arises when the trace component elutes immediately before or after the main constituent. If the trace component appears immediately in front of the main peak then, even though the detector can be operated at a high sensitivity, the limit of detection is determined by the sample size. This is because with increase in sample size the trace peak merges with that of the main component as, for example, in the analysis of traces of acetylene in ethylene on a non-polar column as shown in Fig. 38.

Determination of low concentrations of compounds which elute in the tail of the main component peak is considerably more difficult, and their quantitative measurement poses problems, even for analyses at the percentage level. This applies mainly to those trace components having chemical and physical properties so similar to those of the main component that even the use of special stationary phases does not produce a reversal of the retention times which would enable the trace material to elute before the main component. In this case, it is advisable to separate a fraction containing the trace components from the main component so that the concentration of the latter is significantly reduced. This may be achieved, as already mentioned, by chemical reaction or by a condensation method.

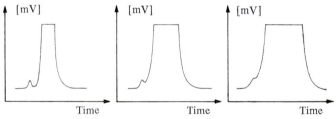

Sample volume: 1–2 ml   Sample volume: 3–5 ml   Sample volume: >5 ml

Fig. 38   Decrease in resolution with increase in sample size.

Another means of increasing the sensitivity of detection for masked trace peaks is to condense them out in a cold trap together with the associated amount of the main component tail, and to subsequently analyse this fraction again on the same column. By repeating this operation several times, it is possible to reach a limit of detection of 0·05 ppm.[12]

Alternatively, the trace materials and main component residue can be left on the column and then, by reversing the carrier gas flow, transferred to a low-temperature column. When this is subsequently heated a further chromatographic separation occurs. The enrichment factor obtained in this way is high enough to produce sufficient quantities of material for spectroscopic identification.[13]

These methods are all unsuitable for rapid quantitative routine measurements such as required, for example, in process control. For these applications the technique of peak-cutting is used. This, of course, does not achieve any enrichment in the usual sense, but for the purpose of analysis it produces a favourable ratio of the amount of trace component to that of the main component. The principle of this peak-cutting method is based on removing the majority of the main component from the column by venting it into the atmosphere. The simplest system is shown in Fig. 39. The method of operation involves gas chromatographic separation on column I followed by venting the bulk of the main component into the atmosphere through the valve *B* via the 4-way valve *A*. The trace component being determined, together with the main component residue, is passed on to column II by rotating *A*. The needle valves *B* and *C* are adjusted so that no fluctuations occur in the carrier gas flow rate which could cause a disturbance of the base line during the

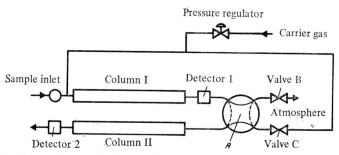

Fig. 39   The simplest method of peak-cutting for removing major components.

rotation of the 4-way valve. This system provides an elegant method of determining trace components which may elute before or after the main component, as shown in Figs. 40, 41 and 42.

In the case when the trace component precedes the main component, the residue of the main peak is vented into the atmosphere by rotating *A*, whilst further separation takes place on column II (Fig. 40).

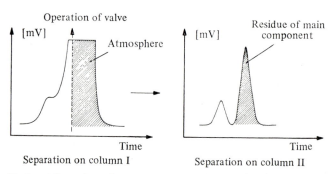

Fig. 40   Peak-cutting when the trace component precedes the main component.

If it is necessary to determine a trace component in the tail of the major peak, then the procedure is carried out in reverse, *i.e.* the first part of the main peak is vented into the atmosphere, and the remainder of the main component is passed along to column II (Fig. 41).

It is also possible to use this technique for the simultaneous determination of trace components which elute both before and after the main component (Fig. 42).

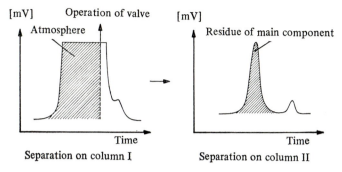

Fig. 41   Peak-cutting when the trace component follows the main peak.

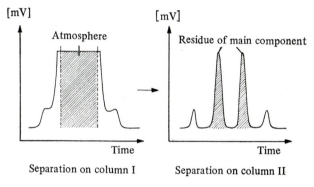

Fig. 42   Peak-cutting when trace components precede and follow the main component.

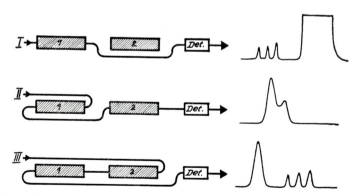

Fig. 43   Column switching with macro-capillary columns (from H. Pauschmann, *Z. Anal. Chem.*, **228**, 39 (1967)).

Pauschmann combined this technique of switching between two macro-capillary columns with back-flushing to determine trace impurities in ethylbenzene.[14] In this way, trace constituents which elute before and after the main component may be detected in one operation (Fig. 43).

First, the compounds which elute before the main component are analysed on column 1 (configuration I). After the detector has indicated the maximum in the main peak (about 96% eluted), the tail fraction containing the trace components is back-flushed and passed on to column 2 (configuration II), where some more of the main component is removed. After back-flushing again from column 2, the final measurement of the trace components, previously masked by the tail of the main peak, can be carried out on column 1 (configuration III).

This peak-cutting procedure has been employed in various forms, including automatic systems, in routine analyses in process gas chromatography. Present developments indicate that in the future this elegant technique will also find widespread application in the laboratory.[15] However, it should be pointed out that calibration may involve some problems since inaccurate 'cutting' of the main peak can simulate trace components which are not present in the sample. Moreover, in order to avoid any changes occurring in the retention times, only very stable stationary phases or solid adsorbents should be used.

References
1. NEUMANN, G. M. and ZIEGLER, A., *J. Chromatog. Sci.*, **7**, 318 (1969).
2. HACHENBERG, H., *Brennstoff-Chem.*, **43**, 225 (1962).
3. JANÁK, J., *Chem. Listy*, **47**, 464, 817, 837 (1953).
4. HORN, O., SCHENK, U. and HACHENBERG, H., *Brennstoff-Chem.*, **38**, 116 (1957).
5. HACHENBERG, H., *Brennstoff-Chem.*, **43**, 225 (1962).
6. HACHENBERG, H. and GUTBERLET, J., *Brennstoff-Chem.*, **49**, 279 (1968).
7. JANÁK, J., *TIPS,* 32 *GC*, Bodenseewerk Perkin Elmer.
8. NOVÁK, J., VAŠÁC, V. and JANÁK, J., *Anal. Chem.*, **37**, 660 (1965).
9. ZHUKHOVITSKII, A. A. and TURKELTAUB, N. M., in *Fortschrittsberichte zur GC*, p. 156, Berlin, Akademie-Verlag (1961).
10. KAISER, R., *Chromatographia*, **1**, 199 (1968).
11. OSTER, H., *Z. Anal. Chem.*, **247**, 257 (1969).
12. PAPA, L. J. and VARON, A., *J. Gas Chromatog.*, **6**, 185 (1968).
13. PRIMAVESI, G. R., *Analyst*, **95**, 242 (1970).
14. PAUSCHMANN, H., *Z. Anal. Chem.*, **228**, 39 (1967).
15. ANON., *Perkin Elmer, Analytical News*, **4** (1970).

1.35. Identification of Trace Components

The techniques described in Section 1.34 are incomplete in that they do not include the identification stage. They are merely concentration techniques which can be used to isolate known trace components or to facilitate the more sensitive and more accurate measurement of the amounts of the trace components. Moreover, the use of chromatographic sorption and reversion gas chromatography is limited to gaseous systems. Many so-called trace analyses are only concerned with the detection of the trace materials and not with their analysis, since the latter term includes the identification of the trace material being determined.

1.351. Retention time and microreactions

Although gas chromatography has developed into the most sensitive and rapid method of separation, when used as a means of identifying a constituent in a mixture, the result, *i.e.* the retention time or retention volume, generally depends on the experimental conditions and is not specific to the compound.

For qualitative interpretation, retention data must be measured on several different polar stationary phases under a variety of experimental conditions. However, this presupposes that the compounds being analysed have already been examined under identical conditions. For routine operation when the qualitative nature of the samples does not vary significantly, the retention method is very useful. However, it is of less use in a research laboratory where a wide variety of mixtures, some of which contain completely new compounds, have to be analysed.

Apart from the fact that this method of identification cannot offer 100% certainty, it presupposes that the corresponding pure compounds are available, and that there is sufficient information about the origin of the sample *i.e.* about possible impurities. One method of verifying these qualitative results, based on measurement of the retention time, is to carry out a specific chemical colour reaction at the exit of the gas chromatograph.[1] These microreactions are frequently of uncertain nature, especially for poorly resolved components.

Chemical pretreatments of samples such as hydrogenation, saponification, esterification, reaction with acids, and extraction, etc. and repeated chromatographing can give very useful information about the nature of the sample. However, final identification of trace components involves spectroscopic methods such as mass spectrometry and infrared, ultraviolet or nuclear magnetic resonance spectroscopy. If the amount of material produced by the concentration procedures described earlier is not sufficient, then this identification must be performed directly at the exit of the gas chromatographic column. For this purpose, even detectors which are only specific to particular classes of compounds can give valuable preliminary information about the actual compounds present (*see* Section 1.312.).

1.352. Specific detection systems: coupling of gas chromatography with other physical methods of analysis

GAS CHROMATOGRAPHY–ATOMIC ABSORPTION SPECTROSCOPY

Atomic absorption spectroscopy represents the most selective and sensitive technique for the detection and quantitative determination of volatile metallic compounds, after they have been separated by gas chromatography. However, the combination is only appropriate when several compounds of a single element have to be determined, since the total amount of an element can be measured directly by atomic absorption spectrometry. For the g.c.–a.a.s. combination, the detector signal is proportional to the amount of the element, and not that of the compound. For quantitative evaluation, a correction factor must therefore be applied to each compound. Kolb *et al.* used this combination to determine lead alkyls in petrols, and reported a limit of detection of 4×10^{-11} g atom Pb/second.[2] Figure 44 illustrates the determination of tetramethyllead (peak *A*) and tetraethyllead (peak *B*) in petrol having a low aromatic content.

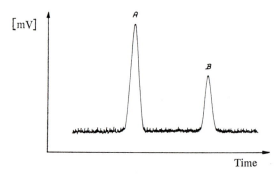

Fig. 44   Determination of lead alkyls in petrol by combined g.c.–a.a.s. (from B. Kolb, G. Kemner, F. H. Schleser and E. Wiedking, *Z. Anal. Chem.*, **221**, 166 (1966)).

Another example of an application of this method is in the analysis of silylated alcohols.[3] By this means alcohols may be selectively analysed, the limit of detection being 0·1 μg Si. Kolb *et al.* considered the small linear range, high noise level for many elements, and also low sensitivity compared with other detectors to be disadvantages of the a.a.s. method.

GAS CHROMATOGRAPHY–INFRARED SPECTROSCOPY

In 1959 Heaton *et al.* had already used a g.c.–i.r. combination to determine traces of hydrocarbons in car exhaust gases.[4] The method involved oxidizing the hydrocarbon over copper oxide to carbon dioxide, which was detected with an infrared spectrophotometer. In this way, for example, even 10 ppm of ethylene gives an easily measurable signal. Although many attempts have been made to develop this combination, owing to the lack of sensitivity of infrared spectrophotometry it cannot be considered as a general method for the identification of trace materials. The following applications will therefore be mentioned only briefly. By using special high-temperature cells directly attached to the gas chromatographic column outlet, infrared spectra may be obtained which enable compounds to be assigned to specific classes of substances.[5]

Bartz and Ruhl employed two specially constructed spectrophotometers in parallel, one of which scanned the region from 2·5 μm to 7 μm in 16 seconds, while the other simultaneously covered the region from 6·5 μm to 16 μm.[6] A special spectrophotometer having a fast scan speed was used by Wilks and Brown in a directly coupled system.[7] Low and Freeman also worked with a fast scanning spectrophotometer which covered the range from 4 μm to 40 μm in one second.[8] The spectra were measured in a flow-through cell at the column outlet over the whole sequence of peaks, and the signals were recorded on magnetic tape. In this way, even unresolved peaks can be characterized. The sensitivity is sufficient to clearly distinguish the major bands for a quantity of material of 0·005 μl. The spectrum obtained with a trace amount of methyl acetate is shown in Fig. 45.

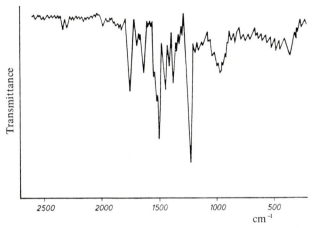

Fig. 45 Identification of trace amounts of methyl acetate with the g.c.–i.r. combination (from M. J. D. Low and S. K. Freeman, *Anal. Chem.*, **39**, 194 (1967)).

Another technique consists of absorbing the components separated gas chromatographically in a low boiling point solvent. This solution is subsequently added to KBr powder and the solvent removed under vacuum, before pressing the powder into a tablet. This method is called the KBr 'eye' technique, and it enables easily interpreted infrared spectra to be obtained with a microgram of material.[9] The method can, of course, be employed only for solids or compounds having very high boiling points. For low boiling compounds, Walz recommends that the vapour of the compound be condensed directly on to a thin film of KBr at the column outlet.[10] Acceptable spectra can be obtained with 3 μl of material.

GAS CHROMATOGRAPHY–MASS SPECTROMETRY[11]

At present the most reliable method of identification is a direct coupling of a mass spectrometer to the gas chromatograph. The two techniques supplement each other ideally because they have approximately equal sensitivities. Separation and identification of individual components in very small samples can therefore be carried out in a single operation without the need for pre-concentration. The advantages of the combination may be summarized as follows:

(*a*) It is the most reliable method of identification since the retention time and the mass spectrum are obtained in a single analysis.

(*b*) The effort and time expended for identification of trace components is considerably less than would otherwise be necessary for the pre-concentration stage.

(*c*) In exceptional cases, the amount of detectable compound is from 10^{-8} to 10^{-11} g s^{-1}. In general, it is not possible to isolate such quantities at the column outlet.

The principal methods of coupling gas chromatographs to mass spectrometers are shown in Fig. 46.[12]

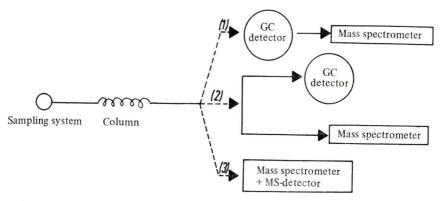

Fig. 46   Methods of coupling a gas chromatograph to a mass spectrometer (from J. A. Völlmin, W. Simon and R. Kaiser, *Z. Anal. Chem.*, **229**, 1 (1967)).

With configurations 1 and 2 there are problems in synchronizing the signals from the gas chromatographic detector and mass spectrometer. Moreover, configuration 1 necessitates a very sensitive detector having a small volume, and which does not destroy the sample. Since such detectors are generally not available—although high sensitivity micro-thermistors might possibly be considered—configurations 2 and especially 3 have so far predominantly been used. The flame ionization detector is used in configuration 2, while in 3 the mass spectrometer itself serves as detector by utilizing the total ion current.

The point at which problems arise in these systems is the connection between the column and ion source. Capillary, support-coated open-tubular (SCOT), and also packed columns may be employed. With the latter, it should be remembered that only a relatively small selection of stable stationary phases can be used. Also, only a limited amount of carrier gas can be admitted into the mass spectrometer. This is generally about 1–2 ml min^{-1}. The carrier gas flow rate used in the chromatograph depends largely on the conditions required for separation, and consequently it is necessary to restrict the flow rate into the mass spectrometer. This is achieved with a splitter system located at the connecting point, and which vents the excess carrier gas before the inlet to the ion source. The dead volume of this splitter should be kept as small as possible to minimize loss of material so that when a difficult separation has been achieved the advantage it is not then lost.

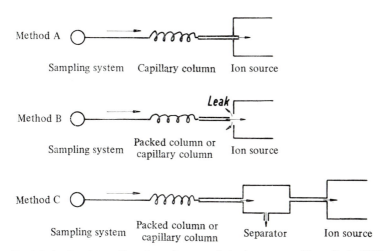

Fig. 47   Methods of coupling the column and the ion source (from J. A. Völlmin, W. Simon and R. Kaiser, *Z. Anal. Chem.*, **229**, 1 (1967)).

The type of connection also determines to what extent air or water vapour penetrate into the ion source by back diffusion, and lead to deterioration in the operating conditions such as *e.g.* instability of the base line and the appearance of a background spectrum.

In recent years several methods have been described for coupling the separating unit to the ion source (Fig. 47), and they will be discussed briefly below.

Method A: the simplest method and that which gives cleanest separation consists of introducing the column directly into the ion source. Although this method of coupling can only be used with capillary columns, it has the advantage that, in addition to preserving the high separation efficiency, no dead volume exists and none of the sample can be lost. Effects due to adsorption or decomposition are largely avoided. According to Völlmin *et al.* the vacuum in the ion source has the possible disadvantage of causing a change in the absolute and relative retention data. However, based on the results of ten Noever de Brauw and Brunnée, and the author's own experience, this observation should not be generalized.[13] *Method A* can only be used with very low carrier gas flow rates. It is therefore not applicable when packed columns are employed, and must be restricted to capillary columns. Although the sample size must then be very small, if a sensitive mass spectrometer is used, it is possible in special cases to achieve limits of detection down to the level of about 10 ppm.[14]

Packed columns necessitate the use of coupling methods *B* or *C*.

Method B: this procedure is known as the leak method and has the advantage of being easily adaptable so that it can be linked with any kind of g.c. apparatus. It therefore has the greatest flexibility. As with *method A*, effects due to adsorption and decomposition rarely occur. The operating conditions for mass spectrometers do not need adaptation to suit gas chromatographic parameters. However, the sensitivity of the leak method to obstructions and temperature fluctuations is a serious practical disadvantage. Moreover, relatively large sample losses and diffusion of atmospheric gases must be taken into account. *Method B* is therefore rarely considered for trace analysis.

The above difficulties may largely be avoided by using molecular separators.

Method C: a molecular separator is located between the column outlet and ion source, and is connected to a pumping system which selectively removes the majority of the carrier gas. In this way, the original pressure of the carrier gas at the column outlet is adjusted to the reduced pressure in the ion source, and simultaneously, enrichment of the trace components is obtained. Völlmin, Simon and Kaiser distinguished four types of separator:

Type I: selective removal of the carrier gas through the wall of a fritted glass tube.[15-18]

Type II: selective removal of the carrier gas through the wall of a Teflon capillary.[19]

Type III: selective removal of the carrier gas by diffusion from the molecular beam.[20-22]

Type IV: selective separation of the eluate from the carrier gas.[23]

The efficiency of separators is characterized by the enrichment factor:

$$K = \frac{x_M}{y_M} \cdot \frac{y}{x} \qquad (10)$$

or by the efficiency:

$$\eta[\%] = \frac{x_M V_M}{x V} \cdot 100 \qquad (11)$$

where:

x_M mole fraction of enriched component at the mass spectrometer inlet
y_M mole fraction of carrier gas at the mass spectrometer inlet

x mole fraction of the component to be enriched at the gas chromatograph outlet
y mole fraction of carrier gas at the gas chromatograph outlet
V flow rate at the separator inlet
V_M flow rate into the ion source.

Depending on the nature of the compounds being analysed, and which of
the above types of separator is being used, enrichment factors can range
from 10 to 10^6 at efficiencies of 20–70%. Molecular separators are therefore
clearly suited to trace analysis since they bring about a concentration of the
component to be identified, and minimize any overloading of the ion source
with carrier gas. When these devices are used, the gas chromatographic
conditions are largely independent of the conditions required in the mass
spectrometer so that it is possible to work with almost optimum conditions
for separation. However, there is the disadvantage that the materials from
which the separator is constructed can cause adsorption and decomposition
effects which may give rise to considerable problems in trace analysis. Nevertheless, these may largely be avoided by silanizing the separator, *e.g.* with
dimethyldichlorosilane.[13] Figure 48 shows two chromatograms obtained while
investigating adsorption effects in the separator.
There are two main types of application of the g.c.–m.s. combination:

I The mass spectrometer may be used as a specific detector for certain
molecular units by focussing on a particular mass number. By this means
only certain quite specific fragment ions are recorded.

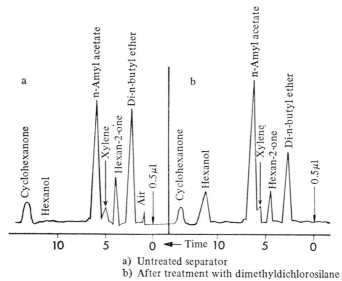

a) Untreated separator
b) After treatment with dimethyldichlorosilane

Fig. 48   Test chromatograms to investigate adsorption in the separator (from
M. C. ten Noever de Brauw and C. Brunnée, *Z. Anal. Chem.*, **229**, 321 (1967)).

II It may be used as a specific detector for each compound by recording the whole mass spectrum in a rapid scanning of the mass range over a few seconds using a high speed photographic recorder.

MASS SPECTROMETER AS A SPECIFIC DETECTOR FOR MOLECULAR UNITS[24]

If an ion beam detector can be focussed on a fixed mass number then it differs from other gas chromatographic detectors in that its specificity can be altered to suit each particular problem. When focussed on a given mass number it will only record peaks belonging to certain compounds. Since the sensitivity of such a system is at least as high as that with a flame ionization detector, then it is possible to determine, for example, 10^{-10} g s^{-1} of benzene, and to use capillary columns. This selective utilization of the mass spectrometer requires prior knowledge of the mass spectrum of the compounds to be

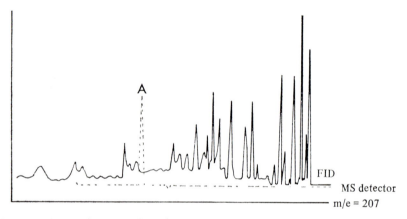

Fig. 49   Selective determination of tetraethyllead in petrol (from D. Henneberg and G. Schomburg, *Z. Anal. Chem.*, **215**, 424 (1966)).

detected. It is desirable to equip the mass spectrometer with a number of collectors, since by recording several characteristic mass numbers considerably more information is obtained. Figure 49 shows the selective determination of 350 vol ppm of tetraethyllead ($m/e = 207$) in petrol (peak *A*). After allowing for the noise level, the method can be used to determine as low as 10 ppm of this compound.

Another related example is the selective determination of alcohols and phenols in the form of their trimethylsilyl ethers.[24]

The advantage of the total ion beam detector compared with the electron capture detector, which is also suitable for this kind of analysis, is found in the greater dynamic range and better stability. In addition, the signal is not influenced by impurities (*e.g.* water) as is often the case with the electron capture detector. A further advantage of detection by means of the total ion beam current compared with the use of other detectors is that the signal due to the carrier gas can be suppressed, so that a better signal-to-noise ratio is obtained.

MASS SPECTROMETER AS A SPECIFIC DETECTOR FOR COMPOUNDS[25]

Only when the mass spectrometer is used to record the whole mass spectrum of a component can it be described as a specific detector for compounds. The majority of instruments used in these combinations are low-resolution mass spectrometers, with which rapid mass scans are carried out. High-resolution double-focussing instruments are generally not suitable for trace analysis since, as with other spectroscopic techniques, attempts to increase the resolution result in a loss of sensitivity.

Instruments with magnetic focussing enable a mass range of m/e from 12 to 500 to be scanned in one second with a resolution normally of 1:600. These relatively high scan speeds can be used even with capillary columns. In the g.c.–m.s. combination, it is usual nowadays to record the total ion beam, rather than use an FID to produce the chromatogram. This can be done in two ways.

The first method is to use a dual ion source in which one section acts as a gas chromatographic ionization detector as shown in Fig. 50.

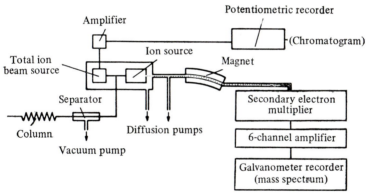

Fig. 50   A g.c.–m.s. combination with dual ion source (from M. C. ten Noever de Brauw and C. Brunnée, *Z. Anal. Chem.*, **229**, 321 (1967)).

In the left hand section of the ion source, which is usually called the total ion beam source or sometimes the manometric ion source, the energy of the electrons is 20 eV. Therefore, only the components being measured are ionized and not the carrier gas (helium). The output signal of this ion source produces the gas chromatogram. A favourable signal-to-noise ratio is obtained since ionization of helium in this section is suppressed. In the right hand section of the dual ion source, the energy of the electrons is 70 eV. The ion beam which gives rise to the actual mass spectra is produced in this chamber. Since the two ion sources are operated independently of each other, the gas chromatogram and individual mass spectra are simultaneously recorded.

The second technique involves introducing the whole sample into a single ion source. A fraction of the total ion beam is intercepted at the orifice in a total ion beam monitor. For this method to attain a sensitivity comparable with that of the flame ionization detector, it must be possible to operate the ion source at 20 eV continuously. In order to obtain a mass spectrum for each analysis comparable with published spectra, it must be possible to adjust the ion source from 20 eV up to higher ionization energies, *e.g.* 70 eV, without defocussing the ion optics. A system of this kind is shown in Fig. 51. When packed columns are employed, a separator can be incorporated at the point where the instruments are coupled.

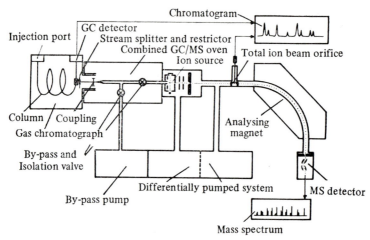

Fig. 51   A g.c.–m.s. combination with one ion source (from Francke, Bodenseewerk, Perkin-Elmer).

A differentially pumped system is required in both forms of instrument employed to record mass spectra which can be analysed for trace components. It removes unionized molecules which have entered the ion source, preventing diffusion into the analysing section of the instrument. Furthermore, the vacuum in the analyser tube must be such that the flight path of the ions is not distorted by collisions with residual gas molecules. The larger the diameter of the tubing in the analyser the more nearly is this achieved. Only by attention to such practical measures is it possible to inject relatively large samples and improve the limits of detection in trace analysis. Figure 52 illustrates the detection of 100 ppm of benzene in hexane by means of the total ion current.

Figure 53 shows a high speed scan of the mass spectrum of benzene in which the mass numbers 78, 77, 52, 51 and 50 are clearly distinguished from the background. The effective amount of benzene corresponds to 5×10^{-11} g since 0·1 μl of a mixture of 100 ppm of benzene in hexane was injected onto the capillary column using a splitting ratio of 1:200. The signal-to-noise ratio for the mass 78 peak is about 100:1. Therefore, by permanently focussing on this peak the minimum amount of benzene which can be detected per unit time is about 3×10^{-13} g s^{-1}.

An elegant method of increasing the sensitivity of detection is to remove the main component at the end of the gas chromatographic column by pumping it out through a stream splitter.[26] This arrangement permits injection of samples up to 10 μl and raises the limit of detection one hundred-fold.

In addition, the use of a short scrubber-column containing a chemically and thermally stable stationary phase can effectively prevent 'bleeding' of the analytical column and contribute to an increase in sensitivity.[27]

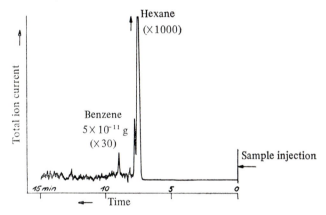

Fig. 52   Detection of 5×10^{-11} g benzene by means of the total ion current (from M. C. ten Noever de Brauw and C. Brunnée, *Z. Anal. Chem.*, **229**, 321 (1967)).

A particularly important example of the use of the g.c.–m.s. combination is the identification of components in the analysis of head space gases, because enrichment procedures generally fail here owing to the low concentration in the vapour phase.[28] Also, the sample size is often so small, *e.g.* in the analysis of commercial samples, that concentration cannot even be considered, and the g.c.–m.s. combination is clearly the sole method offering the possibility of achieving identification.

Precise limits of detection for analysis by means of the g.c.–m.s. combination cannot be specified. They depend on the operating conditions of the gas chromatograph, the performance of the mass spectrometer available, and the type and efficiency of the coupling unit, as well as the nature of the trace components under identification. If it is assumed that an efficient pumping system is connected to the mass spectrometer via a tube of the largest possible

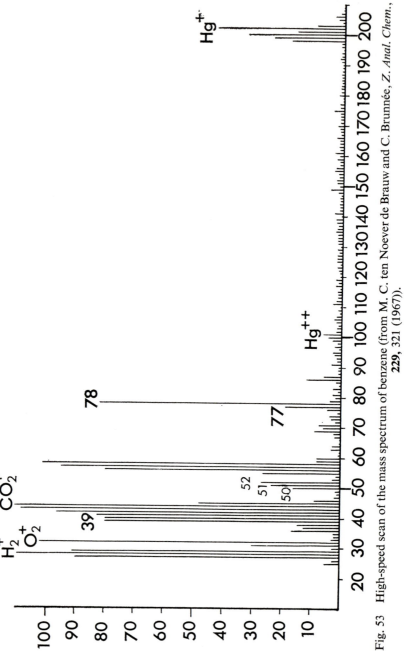

Fig. 53   High-speed scan of the mass spectrum of benzene (from M. C. ten Noever de Brauw and C. Brunnée, *Z. Anal. Chem.*, **229**, 321 (1967)).

diameter, and that the column contains a stable stationary phase, then it is possible to detect 5–10 ppm of hydrocarbons using capillary columns and 100–1000 ppm with packed columns used without a separator.[29] For identification of compounds at even lower concentrations, *i.e.* in the region below 1 ppm, Kaiser recommends a combination of the mass spectrometer with other chromatographic techniques such as multistage g.c., low temperature g.c. using prolonged sampling, and reversion g.c.[30] Kranz has described a coupling system specially designed for trace analyses.[31] To obtain the highest possible detection sensitivity, he operated the ion source at very high ionization currents (up to 300 μA), but it was then necessary to accept a limited life for the indirect filament heater. Tests showed that 10^{-10} g benzene can be detected by this method. Kranz reported that 0·1 ppm of a trace component can be detected in a sample of 1 μl of a mixture, and that the mass spectrum obtained can still clearly be interpreted.

References
1. WALSH, J. T. and MERRIT, C., *Anal. Chem.*, **32**, 1378 (1960).
2. KOLB, B., KEMMNER, G., SCHLESER, F. H. and WIEDKING, E., *Z. Anal. Chem.*, **221**, 166 (1966).
3. MORROW, R. W., DEAN, J. A., SCHULTS, W. D. and GUERIN, M. R., *J. Chromatog. Sci.*, **7**, 572 (1969).
4. HEATON, W. B. and WENTWORTH, J. T., *Anal. Chem.*, **31**, 349 (1959).
5. FEUERBERG, H., MANJOCK, M. and WEIGEL, H., *Z. Anal. Chem.*, **219**, 241 (1966).
6. BARTZ, A. M. and RUHL, H. D., *Anal. Chem.*, **36**, 1892 (1964).
7. WILKS, P. A. and BROWN, R. A., *Anal. Chem.*, **36**, 1896 (1964).
8. LOW, M. J. D. and FREEMAN, S. K., *Anal. Chem.*, **39**, 194 (1967).
9. BERGSTEDT, E. I. M. and WIDMARK, G., *Chromatographia*, **3**, 216 (1970).
10. WALZ, H., Bodenseewerk Perkin Elmer, *TIPS* 13 *GC* (1962).
11. KIENITZ, in *Massenspektrometrie*, Verlag Chemie GmbH., Weinheim/Bergstrasse (1968), section E 1.
12. VÖLLMIN, J. A., SIMON, W. and KAISER, R., *Z. Anal. Chem.*, **229**, 1 (1967).
13. TEN NOEVER DE BRAUW, M. C. and BRUNNÉE, C., *Z. Anal. Chem.*, **229**, 321 (1967).
14. HACHENBERG, H., unpublished work.
15. WATSON, J. T. and BIEMANN, K., *Anal. Chem.*, **36**, 1135 (1964).
16. WATSON, J. T. and BIEMANN, K., *Anal. Chem.*, **37**, 844 (1965).
17. VÖLLMIN, J. A., KRIEMLER, P., OMURA, I., SEIBL, J. and SIMON, W., *Microchem. J.*, **11**, 73 (1966).
18. VÖLLMIN, J. A., OMURA, I., SEIBL, J., GROB, K. and SIMON, W., *Helv. Chim. Acta*, **49**, 1768 (1966).
19. LIPSKY, S. R., HORVATH, C. G. and McMURRAY, W. J., *Anal. Chem.*, **38**, 1585 (1966).
20. BECKER, E. W., BIER, K. and BURGHOFF, H., *Z. Naturforsch.*, **10a**, 565 (1955).
21. RYHAGE, R., *Anal. Chem.*, **36**, 759 (1964).
22. RYHAGE, R., WIKSTROM, S. and WALLER, G. R., *Anal. Chem.*, **37**, 435 (1965).

23. LLEWELLYN, P. and LITTLE-JOHN, D., Pittsburgh Conference on Analytical Chemistry and Applied Spectroscopy, Feb. 1966.
24. HENNEBERG, D. and SCHOMBURG, G., *Z. Anal. Chem.*, **215,** 424 (1966).
25. GOHLKE, R. S., *Anal. Chem.*, **31,** 535 (1959).
26. GOTO, S. and NOSHIRO, M., *Bunseki Kagaku,* **19,** 382 (1970).
27. LEVY, R. L., GESSER, H., HERMAN, T. S. and HOUGEN, F. W., *Anal. Chem.*, **41,** 1480 (1969).
28. HEINS, J. TH., MAARSE, H., TEN NOEVER DE BRAUW, M. C. and WEURMAN, C., *J. Gas Chromatog.*, **4,** 395 (1966).
29. HACHENBERG, H. and GUTBERLET, J., unpublished work.
30. KAISER, R., *Z. Anal. Chem.*, **252,** 119 (1970).
31. KRANZ, R., *Messtechnik*, **6,** 121 (1968).

PART 2

Applications of Trace Analysis

2.1. ANALYSIS OF GASES

2.11. Detection of Traces of Inorganic Gases

2.111. Methods of high sensitivity detection

Whilst traces of hydrocarbons can easily be determined, even in the ppb region, by virtue of the very high sensitivity of the flame ionization detector, for a long time there has been a need for a correspondingly sensitive means of detecting those compounds to which the FID does not respond, or does so with insufficient sensitivity.[1] This class of substance includes almost all of the inorganic gases such as O_2, N_2, CO, CO_2, H_2O, H_2S, SO_2, COS, HCN, NH_3, NO, N_2O and the inert gases, as well as some perchlorinated hydrocarbons, formic acid and formaldehyde. For many years, detection of traces of these materials depended on thermal conductivity detectors, which generally have limits of only 5–10 ppm.

In gas analyses normally carried out at room temperature, thermistor elements have a superior sensitivity. Thus, *e.g.* Engel reported the limits of detection for thermistors to be 10 vol ppm H_2, 5 vol ppm O_2, 5 vol ppm N_2, 5 vol ppm, CH_4, 5 vol ppm CO_2, 10 vol ppm CO, and 5 vol ppm SO_2.[2]

Trace analyses of pure gases, particularly monomers such as ethylene or propylene will therefore always be incomplete if the inorganic components such as H_2, O_2, CO, CO_2, etc. cannot be determined with the same sensitivity as the organic impurities. The following possible methods exist for improving the sensitivity of detection of gaseous inorganic trace components:

(*a*) Increase of sample size

(*b*) Pre-concentration techniques (*see* 1.34)

(*c*) Reaction gas chromatography, *i.e.* conversion into compounds which can be detected with the flame ionization detector. Examples are the hydrogenation of CO and CO_2 to give methane, or the reaction of water with calcium carbide to form acetylene.

(*d*) Increase in the sensitivity of thermal conductivity cells

(*e*) Use of ionization detectors and detectors which are specific to certain groups of compounds.

The *use of large samples* is only possible when trace components can be well resolved from the main constituent, and this is not always the case.

Apart from this difficulty, the technique gives only a relatively small increase in sensitivity. A further problem associated with the use of large samples together with thermal conductivity detectors arises in the determination of components which elute relatively rapidly, such as hydrogen. The 'start-peak', which is attributed to the pressure surge on introducing a large sample, can render such analyses impossible. Engel eliminated this problem by raising the pressure in the injection port.[2] This causes the start-peak to become smaller, but if the pressure is increased too much the signal becomes a negative deflection. However, correct adjustment of the pressure can compensate completely for this effect, the pressure of the carrier gas before the column being the critical factor.

In the analysis of gases, the *pre-concentration technique* can employ Janák's method of preparative gas chromatography, or utilize the reactivity of the main component. Thus, for example, ethylene may be almost quantitatively absorbed in large nitrometers filled with bromine or sulphuric acid–silver sulphate solutions, while the inert gaseous constituents, which have been concentrated tenfold, can be determined (*see* 1.341). Apart from the fact that such procedures are very time-consuming, undesirable side reactions sometimes occur giving an inaccurate analytical result.

For inert gases, pre-concentration methods, which must then be carried out at very low temperatures, are also often successful. Thus, Zocchi determined 8 ppb of methane in helium, hydrogen and neon.[3] Concentration was carried out at $-214°C$, and an enrichment of 100% could be attained using a low temperature trap filled with silica gel and a molecular sieve. Since the methods described above are very time-consuming, direct determination of trace components by a sensitive method of detection is to be preferred.

One possibility consists of converting inorganic gases and vapours into compounds which can be detected by the flame ionization detector, *i.e.* by *reaction gas chromatography*. This method has been successfully used to determine <1 ppm of CO and CO_2 in ethylene by hydrogenating them over nickel at 250–300°C to give methane.[4]

$$CO + 3H_2 \xrightarrow[\text{Ni}]{250°C} CH_4 + H_2O$$

$$CO_2 + 4H_2 \xrightarrow[\text{Ni}]{300°C} CH_4 + 2H_2O$$

Since the hydrogenations of both CO and CO_2 must be carried out at specific temperatures, the danger exists that at 300°C the following reaction can also take place in the case of the CO:[5]

$$2CO \longrightarrow CO_2 + C$$

However, it is possible to work at 300°C if the reaction is carried out in the presence of excess hydrogen, and this is ensured by using hydrogen as the carrier gas. Comparative experiments showed that under these conditions equal amounts of CO, CO_2 and CH_4 gave equal peak areas, within experimental error, thus demonstrating that hydrogenation of CO and CO_2 to CH_4 proceeds in a quantitative manner. If the reaction is performed at 250°C, then CO_2 is not quantitatively hydrogenated. Moreover, there must be no cold spots present on the nickel hydrogenation catalyst because CO may then be eliminated from the determination (carbonyl formation!). Figure 54 shows the experimental arrangement used.

The method of using the apparatus can be understood from Fig. 54. The catalyst chamber is located between the chromatographic column and the FID, and consists of a quartz tube packed with Sterchamol coated with 10% Ni freshly prepared by reducing Ni $(NO_3)_2$. Activated charcoal is used as the column packing for the separation of CO, CH_4, CO_2 and C_2H_4.

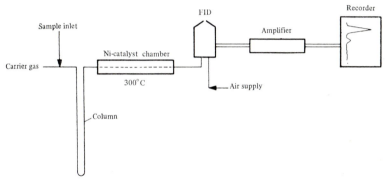

Fig. 54   Experimental arrangement for the gas chromatographic determination of CO and CO_2 with the FID.

Figure 55 shows the chromatogram of a commercial grade of ethylene. In addition to methane and ethylene (as C_2H_6), CO and CO_2 appear as CH_4 based on their usual retention times on the activated charcoal column.

It may be concluded from Fig. 55 that 100 ppb CO and CO_2 can be determined in ethylene. Smaller amounts cannot be detected because the hydrogenation unit located before the detector gives rise to a relatively large base line fluctuation.

Porter and Volman carried out the same conversion of CO to methane.[6] They gave a detailed discussion of the problem of peak tailing and its effect on the elution of carbon monoxide, and stated that 260°C is the best elution temperature.

The reaction described above also enables traces of hydrogen to be measured as methane. Moreover, oxygen may also be successfully determined in the form of methane with the flame ionization detector by carrying out two conversion reactions.[7] The first step is the formation of CO (platinized activated charcoal at 900°C), and in the second stage this is converted to methane as already described. This method has a limit of detection of 0·4 ppm for oxygen.

Finally, the formation from non-volatile radioactive materials of volatile compounds, which can be measured with a Geiger counter, can be used for the determination of traces of inorganic components such as bromine, fluorine, chlorine or nitrosyl chloride.[8]

These indirect methods of determining traces of substances using chemical reactions are, however, frequently rather inconvenient and require constant attention. They are therefore now mostly of historical interest only. The most

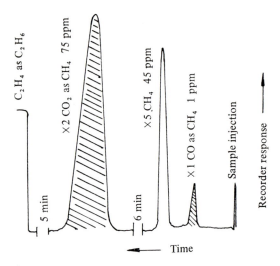

Fig. 55   Determination of traces of CO and CO_2 in C_2H_4 with the FID. Analytical conditions: column: charcoal (Supersorbon, activity 3); column diameter: 6 mm; column length: 1 m; carrier gas: H_2 ($3\cdot5\,l\,h^{-1}$); combustible gas: H_2 ($3\cdot5\,l\,h^{-1}$); air: $25\,l\,h^{-1}$; recorder: 5 mV; electrode voltage: 130 V; working resistance: 40 GΩ; Ni catalyst length: 350 mm; diameter: 6 mm; hydrogenation temperature: 300°C.

reliable method of detecting traces of inorganic gases is by the direct method with a suitable detector. First, ways will therefore be considered of *increasing the sensitivity of thermal conductivity detectors.* Attempts to operate twin filament cells at high currents (> 300 mA) usually fail owing to the sensitivity of these systems to small fluctuations in flow rate, pressure and temperature.

Another possible way of increasing the sensitivity would be to electronically amplify the signal produced by the detector. However, the sensitivity of detection is not improved by this means since the background noise of the measuring unit is amplified to the same extent. Moreover, since the fluctuation of the base line with time is almost the same as the geometric form of the signal, no useful effect is obtained by electrical filtering.[9]

An approximately sixfold increase in sensitivity can be obtained without increasing the noise level by cooling thermal conductivity detectors.[10] Hachenberg and Gutberlet showed that it is possible to detect down to <1 ppm of inorganic gases with cooled thermistor detectors.[11] The definition of the sensitivity of such detectors has already been discussed (*see* 1.31). It is given by the S-value, which is independent of recorder sensitivity, carrier gas flow rate, and chart speed:[12]

where
$$S = \frac{AC_1C_2C_3}{W} \ (\text{mV ml mg}^{-1}) \tag{12}$$

A   peak area (cm^2)
C_1 recorder sensitivity (mV cm^{-1})
C_2 reciprocal chart speed (min cm^{-1})
C_3 carrier gas flow rate (ml min^{-1}) (corrected)
W   mass of the substance (mg).

The value of S is also given by the relation:

$$S \simeq \left(\frac{Ri}{T_a}\right)^2 \simeq \left(\frac{U}{T_a}\right)^2 \tag{13}$$

where R is the resistance of the thermistor, i is the current through the thermistor, and T_a is the ambient temperature (equal to that of the detector block).

Consequently, an increase in the sensitivity of a thermistor conductivity cell should be obtained in the following ways:

(a) increase in i or U
(b) increase in R
(c) lowering of T_a.

Increases in U or i are controlled by the characteristic curves of the thermistors.[13] Increases in R are limited for the same reason, since as R is raised the linear region of the curve becomes progressively smaller. Moreover, at high values of R the temperature of the detector block must be accurately controlled, since even very small differences in the measuring unit temperature can disturb its equilibrium state.

It follows that a practicable method of increasing the sensitivity is to reduce T_a i.e. by cooling the thermistor cell. Figure 56 shows this increase in the sensitivity of a thermistor detector to methane at different temperatures and operating voltages.[11]

It is evident from Fig. 56 that choice of the correct voltage is extremely important if the maximum sensitivity is to be attained at temperatures below 25°C. This is because of the marked temperature dependence of the thermistor characteristics. As a result of the steep rise of the curves, voltage must be measured to the second decimal place in order to remain within the region of optimum sensitivity at each particular temperature. It also appears that with the thermistor detector employed no useful increase in sensitivity is achieved at temperatures below 15°C. At 10°C, measurements could only be made at detector sensitivities up to 60 000 mV ml mg^{-1}. Above this point, the noise level became too great (broken line in Fig. 56). Below 5°C the sensitivity again decreases sharply.

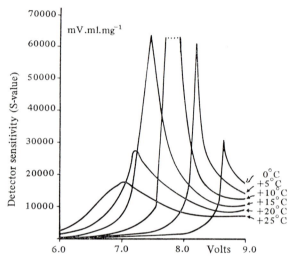

Fig. 56   Variation with operating voltage of the sensitivity of a thermistor detector to methane at different temperatures. Experimental conditions: carrier gas flow rate: 3 l He/h; column packing: 5 Å molecular sieve; column length: 2 m; internal column diameter: 4 mm; column temperature: 22°C; sample size: 0·2577 ml; recorder span: 1 mV.

On a purely theoretical basis, it is assumed from equation 13 that the sensitivity would increase with further reduction in temperature. However, equation 13 is really valid only for a static system. Limitations arise in a dynamic situation, such as in gas chromatography.The thermistor is then no longer under conditions where it can record the undistorted peak shape owing to the limitations of the time constant, which at 25°C can be one second. At lower temperatures or with a greater resistance, the time constant increases considerably. This can be avoided by working at very low carrier gas flow rates. However, for practical gas chromatographic analysis, particularly trace analysis, there are limitations since the low flow rates will result in retention times which are too long, and consequently bring about a reduction in sensitivity owing to the production of flat peaks. The increase in sensitivity obtained by lowering the temperature would therefore be partly annulled.

Figure 57 shows the determination of 2 vol ppm O_2 and 40 vol ppm N_2 in ethylene with a cooled thermistor detector.[11]

From Fig. 57 it can be seen that, with a noise level of 1.2×10^{-2} mV, 0.5 vol ppm O_2 can still be determined accurately.

The use of cooled thermistor cells is not very satisfactory for extended periods or for routine operation since there is considerable difficulty in controlling the noise level and drift associated with this technique.

Considering the present state of development of detectors, *radiation ionization detectors* (*see* 1.31) are to be preferred to all other possible types for the analysis of traces of inorganic gases. The first detector of this kind to be successfully used for this purpose was the argon ionization detector.[14]

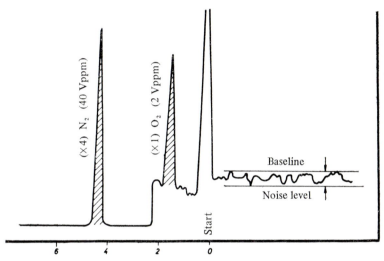

Fig. 57   Determination of 2 vol ppm O_2 and 40 vol ppm N_2 in gases using a 2 m molecular sieve column and a cooled thermistor detector (15°C).

It is based on the principle that the argon carrier gas is ionized by α- or β-radiation to produce free electrons. An electric field accelerates these electrons to an energy which is sufficiently high to convert further argon atoms into an excited metastable state (11.7 eV). These argon atoms transfer their energy to molecules of the organic compounds separated gas chromatographically, so that these are now ionized. The resulting positive ions generate an ionization current in the electric field.[15] This process can only take place with substances having an ionization potential less than 11.7 eV. These are mainly organic compounds.

Molecules having ionization energies greater than the 11.7 eV excitation energy of argon are not ionized by the metastable excited argon atoms. These are the inorganic gases such as O_2, N_2, H_2, CO_2, CO, etc., which can be detected by this method only if they quench the excitation energy of the argon. This effect can be enhanced by mixing a fixed proportion of the vapour of an organic compound with the argon carrier gas before it enters the detector. The organic molecules are then efficiently ionized by energy transfer from the excited argon atoms, thus giving rise to a considerable ionization current. On the other hand, inorganic components will quench part of the excitation energy of the argon atoms before they are able to ionize the organic molecules. Under suitable operating conditions, this leads to such a marked diminution of the ionization current that a very sensitive detection of inorganic gases becomes possible.

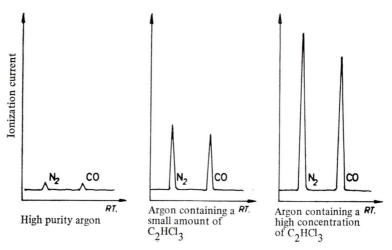

Fig. 58   Variation in peak height for the same quantities of gases as a function of argon purity (from R. Lesser, *Angew. Chem.*, **72**, 775 (1960)).

Thus, *e.g.* Willis added about 1 ppm of ethylene to the carrier gas before it entered the detector and attained a limit of detection of about 0·5 ppm for H_2, O_2 and CH_4.[16] Identical results were obtained on adding the same amount of acetylene. Lesser also made a detailed investigation of this technique.[17] Figure 58 shows the enhanced sensitivity towards small amounts of N_2 and CO when increasing quantities of trichloroethylene vapour are added.

According to Lesser, it is difficult to maintain the concentration of the organic additive constant for long periods. This problem may be overcome by placing a narrow glass tube filled with a solid organic compound, *e.g.* 1,2,3,4-tetrachlorobenzene in front of the detector. At constant temperature, the concentration of the organic compound introduced into the argon remains constant and corresponds to its vapour pressure. In this way, Lesser was able to detect 4 ppm H_2, 6 ppm N_2 and 13 ppm O_2 in niobium.

Bothe and Leonhardt have described a device which enables low concentrations of propane to be reproducibly added to argon.[18] It was shown that

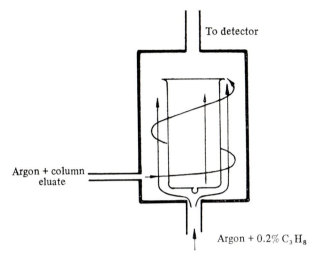

Fig. 59   Gas mixing cell (from H.-K. Bothe and J. Leonhardt, *J. Chromatog.*, **19**, 4 (1965)).

the optimum concentration of propane is about 0·1% for the detection of inorganic gases. To achieve this, a mixture of 0·2% propane in argon is introduced into the carrier gas in a ratio of 1:1 in a specially designed mixing cell. The construction of this mixing unit, which is placed between the column and detector, is shown in Fig. 59.

Bothe and Leonhardt found that for inorganic materials the sensitivity of the argon ionization detector depends on the electric field strength in the

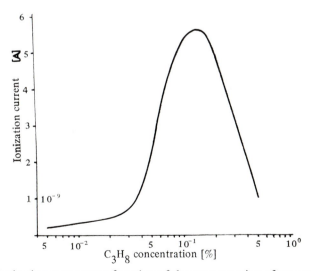

Fig. 60   Ionization current as a function of the concentration of propane in argon. (from H.-K. Bothe and J. Leonhardt, *J. Chromatog.*, **19**, 5 (1965)).

detector, and on the concentration of the organic compound added to the argon.

Figure 60 shows the ionization current in the detector as a function of the concentration of propane in argon.

The ionization current increases linearly up to a certain concentration of propane. This is the region in which the detector is used for the quantitative determination of organic compounds. Beyond this point, the ionization current rises considerably more rapidly and reaches a maximum. At very high field strengths electrical breakdown occurs. At even higher propane concentrations the ionization current falls again. In the presence of an inorganic compound, the ionization current is reduced but the shape of the curve remains unchanged.

It may also be deduced from the graph that the concentration of the organic constituent in argon must be so chosen that small variations have no effect or only a very small one, on the background ionization current. This requirement is met with concentrations of 0·1–0·15% propane, which correspond to the maximum of the curve. Bothe and Leonhardt give the following values for the limit of detection

Substance	Limit of detection
He	$7·40 \times 10^{-8}$
Kr	$1·56 \times 10^{-6}$
H_2	$3·76 \times 10^{-10}$
O_2	$2·98 \times 10^{-9}$
N_2	$3·93 \times 10^{-9}$
CO	$2·61 \times 10^{-9}$
CO_2	$4·15 \times 10^{-10}$
CH_4	$8·05 \times 10^{-10}$

These workers have solved the problem of simultaneously detecting traces of inorganic and organic components by means of a dual-cell argon ionization detector. After gas chromatographic separation, the trace components are passed through the two cells in series. In the first detector, which is operated without introducing propane, the organic components are detected, whilst the traces of inorganic gases are determined in the second detector which is flushed with propane. A glass sinter inserted between the two cells prevents the organic compound added to the carrier gas reaching the first detector. In this way, a chromatogram is obtained which simultaneously displays the peaks resulting from traces of inorganic and organic constituents. Figure 61 shows a typical analysis of 10 μl of coal gas.

Another method of increasing the sensitivity of the argon detector towards permanent gases is to use helium as the carrier gas. In 1958, Lovelock had already referred to this method of using the β-radiation argon detector.[14] Lipsky *et al.* found that, with purest helium, the sensitivity of this radiation ionization detector is highly satisfactory for inert gases and, in particular, the response is approximately equal for all substances.[15] This is because of the high energy (19·8 eV) of the metastable helium atoms compared with all other materials.

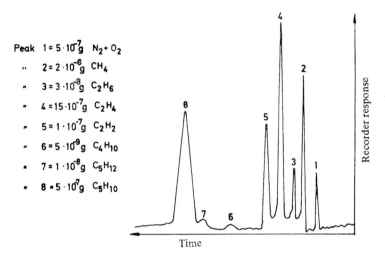

Peak 1 = $5 \cdot 10^{-7}$ g $N_2 + O_2$

" 2 = $2 \cdot 10^{-6}$ g CH_4

" 3 = $3 \cdot 10^{-8}$ g C_2H_6

" 4 = $15 \cdot 10^{-7}$ g C_2H_4

" 5 = $1 \cdot 10^{-7}$ g C_2H_2

" 6 = $5 \cdot 10^{-9}$ g C_4H_{10}

" 7 = $1 \cdot 10^{-8}$ g C_5H_{12}

" 8 = $5 \cdot 10^{-7}$ g C_5H_{10}

Fig. 61   Gas chromatogram of a sample of coal gas (from H.-K. Bothe and J. Leonhardt, *J. Chromatog.*, **19**, 10 (1965)).

In recent years, there have been a number of further reports concerning the use of helium as the carrier gas, and those of Berry and of Bourke *et al.* deserve special mention.[19-23] Berry reported a detection limit of 10^{-12} mole of component/ml of carrier gas, but this can be attained only with extremely pure helium (99·999%).

The purest grade of helium available commercially must therefore be further purified. Molecular sieve at $-196°C$, titanium at $1000°C$ and Hopcalite at $400°C$ are used for this purpose. Bourke *et al.* carried out a suitable purification in four steps:

(*a*) Molecular sieve at $15°C$ and $-196°C$ for the removal of water, carbon dioxide and other easily condensable gases.

(*b*) Titanium at $800°C$ to remove nitrogen.

(*c*) Hopcalite, a mixture of copper and manganese and their oxides, to remove oxygen and oxidize hydrogen to water and carbon monoxide to carbon dioxide.

(*d*) Molecular sieve at 15°C and −196°C to remove water and carbon dioxide formed in (*c*).

The helium flow rate in this purification process should be between 100 and 500 ml min^{-1}.

With an ionization current between 1.59×10^9 A and 3.00×10^9 A there is a linear response for CO_2, CO, O_2, Ar, N_2 and H_2 over the concentration range 0–14 vol ppm. Neon is an exception since its ionization potential (21 eV) is higher than the metastable excited state of helium (19·8 eV), and it therefore produces a negative signal.

Gnauck has used neon as the carrier gas instead of helium.[24] It is then necessary for the ionization chamber of the detector to be relatively large, since even at low voltages (200 V) discharges occur which lead to instability. Also when using tritium as the radiation source, a higher sensitivity was obtained than with strontium-90 as the emitter. With a sample size of 10 ml this system can determine, for example, 2.5×10^{-8} g of N_2 in argon.

This method of using the β-radiation argon ionization detector with helium as carrier gas may be regarded as the first step towards the modern *helium ionization detector*. This was developed to its present efficiency for the detection of ppb quantities by Hartmann and Dimick, in 1965, and it is now considered to be the most sensitive g.c. detector.[25] The initial failures of the detector were mostly due not to the measuring system itself but to very small leaks in the gas chromatograph. Particular attention must be paid to the sampling valve. Conventional designs, which have sliding surfaces in the valve, can only be used if the whole gas sampling system is flushed with helium so that no spurious components from the atmosphere can enter the sample being analysed.

Poy and Verga have developed a special sampling system for the helium detector which has no sliding or rubbing surfaces and involves only diaphragm valves.[26] Moreover, the whole system is isolated from the atmosphere by a protective helium blanket. The specifications given by the authors are as follows:

Diffusion of atmospheric air: <0·1 ppm;
Diffusion of sample gas into the carrier gas: <0·01 ppm;
Ghost peaks of oxygen and nitrogen during each filling of the sample valve: <0·02 ppm;
Reproducibility of sampling: ±0·1%.

The efficiency of this sampling system, which can also be operated automatically, was tested with more than 100 000 sample injections. It should be pointed out that sample collection is equally important and should also be carried out under carefully controlled conditions (*see* 1.331).

As already mentioned, careful attention must be given to the purification of the helium, and in particular the 5 Å molecular sieve used must be kept at liquid nitrogen temperature. After being used for about eight hours, the molecular sieve must be heated at 300°C in a stream of helium.

Poy and Verga recommend using either helium of 99·999% purity or purifying the normal grade of helium by both of the following methods:

The cylinder gas is first passed through a leak-free pressure reducing valve and treated with 5 Å molecular sieve at 20°C and −196°C to remove the bulk of the impurities. The molecular sieve is subsequently reactivated at 400°C. Further purification is carried out over titanium sponge at 800–1000°C, and over Hopcalite at 300–400°C. Then it is again passed through 5 Å molecular sieve at 20°C and −196°C to remove water vapour resulting from oxidation of hydrogen by the Hopcalite.

In addition to this chemical purification of the helium, physical treatment is recommended in which the helium is passed at high pressure and temperature through a diffusion chamber incorporating a large number of quartz capillary tubes. The rate of diffusion of helium through the walls of these capillaries is greater than that of other gases, and helium with an impurity content of less than 0·5 ppm is obtained by this method. Besides these purification procedures, which are rather troublesome and complicated for routine operation, Poy and Verga recommend using commercial helium of the normal purity, from which only the water is removed by means of P_2O_5. The residual impurity level of the carrier gas is then in the ppb region so that, in the analysis of traces of oxygen, nitrogen and argon, the helium detector exhibits the familiar peak inversion. Because of this the authors have developed a method in which the purity of the helium is reduced by adding small amounts of hydrogen in the range 5–15 ppm. This added hydrogen is maintained at a constant level by means of a PTFE capillary tube and causes the detector signal from the above gases to revert to a positive value which is retained as long as the impurity level is less than that of the carrier gas.

In addition to the problem of gas leaks and this change in helium composition, the high sensitivity of the helium detector to water vapour, entrapped during each sample measurement, must be taken into account in trace analysis (*see* 1.311). If these three sources of interference are eliminated, trace amounts well below 1 ppm may routinely be determined without serious difficulty. Thus, for example, the following limits of detection may be attained[27]:

O_2	4 ppb
N_2	2 ppb
N_2O	11 ppb
CH_4	17 ppb
C_2H_6	7 ppb
C_3H_8	11 ppb
SO_2	75 ppb
CO	2 ppb

The linear range is entirely adequate for the concentration range in which this detector is normally employed (0·01–5 ppm) as Fig. 62 shows for the case of O_2, CO and N_2O.

One practical disadvantage of the helium detector is the limited choice of suitable column packings since, as with the FID, even minimal 'bleeding' of the column produces a relatively high background current. Because of this, only columns packed with adsorbents such as aluminium oxide, silica gel, activated charcoal and molecular sieve, and also the various types of Porapak, have been employed so far. Even with this last material, the problem can

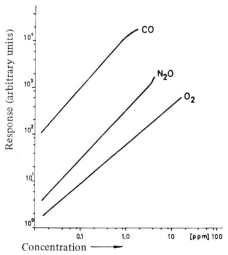

Fig. 62   Linear range of the helium detector for CO, O_2 and N_2O in the region 0·01–10 ppm (from Varian Aerograph, manufacturer's publication 10/1968/550, A-1007 (1968)).

still occur owing to removal of residual monomers and oligomers. Capillary columns also give rise to difficulties on account of the low carrier gas flow rate, since to obtain a useful sensitivity the detector needs a certain amount of helium. Figure 63 shows the routine determination of the traces of inorganic constituents in a commercial grade of ethylene. This was performed simultaneously with two helium detectors connected to two different columns, 5 Å molecular sieve and Porapak Q.[11] The concentrations of CO and N_2O are less than 1 ppm.

The high sensitivity of this detector also permits analysis of very small samples. This results in significantly better resolution factors for pairs of substances which are difficult to resolve, e.g. oxygen and argon. Separation can even be carried out at room temperature on a 2 m long 5 Å molecular sieve column using a sample volume of 0·2 ml (Fig. 64).

Another highly sensitive detector suitable for the detection of traces of inorganic and inert gases is the *radiofrequency discharge detector*, which is mentioned here for the sake of completeness. It is based on the principle that, when a radio-frequency supply of about 10 kV is applied to an ionization chamber at atmospheric pressure, the electric field set up creates a glow discharge.[28] The substances present in the carrier gas cause changes in the luminous intensity and colour which are detected by a photocell and amplified.

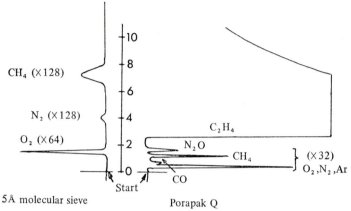

Fig. 63   Trace analysis of pure ethylene using the helium detector and 2 m long 5 Å molecular sieve and 2 m long Porapak Q columns at 22°C; sample size: 0·26 ml.

This technique can also easily detect 0·1 ppm of inert gases.[11] Figure 65 shows a comparison with a high-sensitivity thermal conductivity cell.

The disadvantage of the highly sensitive radio-frequency discharge detector is its limited flexibility, since there are many problems in trace analysis to which it cannot be adapted. Its performance is excellent for the analysis of

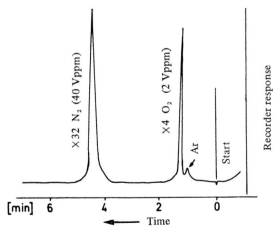

Fig. 64   Determination of 2 vol ppm O_2 and 40 vol ppm N_2 in gases with a 2 m long 5 Å molecular sieve column and a helium detector.

traces of inorganic gases and gaseous saturated hydrocarbons, but on attempting to analyse polymerizable materials such as ethylene or propylene, deposits are formed on the walls of the detector chamber causing the unit to become very insensitive or temporarily unusable. The cell can be made to recover by repeated feeding with oxygen.[11] The following comparison of sensitivities for the detection of inert gases by the detectors discussed above may be made on the basis of measurements carried out by Hachenberg and Gutberlet[11]: If the sensitivity of a single-filament thermal conductivity detector is unity (Gow-Mac, Pretzel, W2, 270 mA) then by comparison the

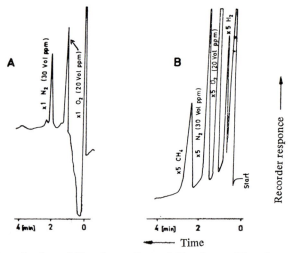

Fig. 65   Determination of 20 vol ppm O_2 and 30 vol ppm N_2 using a thermal conductivity detector (Gow-Mac-W2, bridge current 270 mA, column 1 m long molecular sieve) (*A*), and a radio-frequency detector (*B*).

double filament thermal conductivity detector (Gow-Mac, W2X, 270 mA) is about 10 times more sensitive, the cooled thermistor detector (Perkin-Elmer, F 7/T, +15°C, 7·40 V) about 40 times, the radio-frequency discharge detector (Lambert AL5, Societé l'Air Liquide) about 300 times, and the helium detector (Varian Aerograph 1932-B) about 500 times more sensitive.

2.112. Methods of separation

Although the actual number of inorganic gases to be analysed is limited compared with other gas chromatographic analyses, such as that of hydrocarbons, new and difficult problems of resolution are always arising, especially in trace analysis. An example of this is the determination of traces of argon and oxygen in each other. The separation of these two gases is relatively simple as they are present in comparable concentrations. However, it is difficult to measure ppm amounts of one in the other. Argon may be determined by using hydrogen as the carrier gas and removing the oxygen by means of palladium.[29] Pyrogallol can also be used for this purpose. Both of these methods are differential techniques and require analysis to be carried out twice. They are relatively inaccurate compared with direct methods, and require too much time for routine analyses.

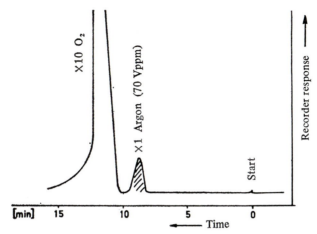

Fig. 66   Determination of traces of argon in oxygen on a 2 m molecular sieve column at $-80°C$.

Determination of traces of argon in oxygen is therefore preferably carried out directly on a 2 m long 5 Å molecular sieve column at $-80°C$ (Fig. 66).
  The converse problem, *determination of traces of oxygen in argon*, does not arise so frequently but it is nevertheless just as important when argon is

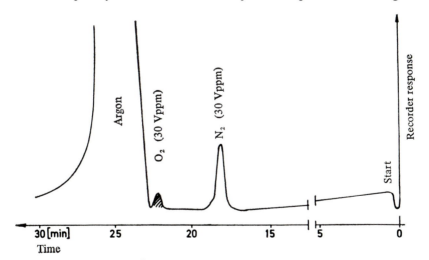

Fig. 67   Determination of traces of oxygen in argon with a Porapak Q column at $-60°C$.

TABLE 2 Separation of various inorganic gases on **21 columns** (from R. M. Bethea and M. C. Meador, *J. Chromatog. Sci.,* **7**, 655 (1969))

Column	Peaks in order of elution								
	1	2	3	4	5	6	7	8	9
1 3 m Silicone oil SF-96 on Chromosorb T	Air, NO	CO_2, N_2O	NH_3	H_2S	SO_2	Cl_2	—	—	—
2 3 m Triacetin on Chromosorb T	Air, NO	CO_2, N_2O	H_2S	NH_3	Cl_2	SO_2	—	—	—
3 3 m Di-n-decyl phthalate on Chromosorb T	Air, NO	NH_3	CO_2, N_2O	H_2S	Cl_2	HCl	SO_2	—	—
4 3 m Arochlor 1232 on Chromosorb T	Air, NO	CO_2, N_2O	NH_3	H_2S, NO_2, HCl	SO_2, Cl_2	—	—	—	—
5 3 m QF-1 (FS-1265) Fluoro on Chromosorb T	Air, NO	CO_2	N_2O, HCl, H_2S	Cl_2	NO_2	NH_3	SO_2	—	—
6 3 m Silicone XE-60 (nitrile) on Chromosorb T	Air, NO, N_2O, CO_2	H_2S, NH_3	Cl_2	HCl, SO_2,	NO_2	—	—	—	—
7 3 m Fluorolube HG 1200 grease on Chromosorb W, AW, DMCS	Air, NO, CO_2, N_2O, H_2S, HCl	SO_2, Cl_2	NO_2	NH_3	—	—	—	—	
8 3 m Kel F90 grease on Chromosorb W, AW, DMCS	Air, NO, CO_2, N_2O, H_2S, HCl	SO_2, Cl_2	NO_2	—	—	—			

No.	Column	1	2	3	4	5	6	7	8	9
9	3 m Halocarbon 11-14 on Chromosorb W, AW, DMCS	Air, NO, N_2O, H_2S, HCl	CO_2	SO_2	Cl_2	—	—	—	—	—
10	3 m Silicone DC 200 oil on Chromosorb W, AW, DMCS	Air, NO, CO_2, N_2O	H_2S	SO_2, HCl, NH_3	NO_2	Cl_2	—	—	—	—
11	1·8 m H_3PO_4 on Porapak Q	Air, NO, NO_2, NH_3	CO_2	N_2O	Cl_2	H_2S, HCl	SO_2	—	—	—
12	1·8 m Polypak 1	Air	NO, SO_2	NO_2	—	—	—	—	—	—
13	1 m Polypak 1	Air, NO	CO_2	N_2O	NH_3	HCl	NO_2	H_2S	Cl_2	SO_2
14	3 m Di-n-ethylhexyl adipate on Chromosorb W, AW	Air, NO, CO_2, N_2O	NH_3, H_2S	Cl_2	SO_2	HCl	NO_2	—	—	—
15	1·8 m Porapak Q	Air	CO_2	SO_2, NO_2	N_2O	—	—	—	—	—
16	1·8 m Porapak QS	Air, NO	CO_2	NH_3, N_2O	HCl, Cl_2	H_2S, NO_2	SO_2	—	—	—
17	1·8 m Porasil B	Air, CO_2, N_2O, NO, H_2S	—	—	—	—	—	—	—	—
18	2·5 m Arochlor 1232 on Porapak Q	Air, NO	CO_2, N_2O	H_2S, HCl	SO_2	Cl_2	NO_2	—	—	—
19	2·5 m Arochlor 1232 on Porapak R	Air, NO	CO_2, N_2O	NO_2, H_2S, NH_3	SO_2	Cl_2	—	—	—	—
20	2·5 m Arochlor 1232 on Porapak Q + 2·5 m Arochlor 1232 on Porapak R	Air, CO_2, NO	NO	N_2O	NO_2, HCl, Cl_2	SO_2, NO_2	—	—	—	—
21	2·5 m Porapak Q	Air, NO	CO_2, N_2O	N_2O, NH_3, H_2S	NO_2, HCl, Cl_2	SO_2	—	—	—	—

used as purging gas for certain processes. This analysis has always been difficult by direct gas chromatography because the oxygen appears in the tail of the argon peak with molecular sieve columns. Even with the optimum separation of these two components, the limit of detection of oxygen is only 100–200 vol ppm. As in other trace determinations, it is also desirable here to elute the trace before the main component. This may be achieved with Porapak Q at $-60°C$ as shown in Fig. 67.

The problems involved in separating a small number of inorganic gases are illustrated by the work of Bethea and Meador on the determination of traces of NO, NO_2, N_2O, SO_2, HCl, H_2S, Cl_2, NH_3 and CO_2 in the range 1–50 ppm.[30] From the extensive literature in this field, these workers selected 21 columns enabling quantitative analyses to be performed at levels below 100 ppm.[31–40] (Table 2).

Obermiller and Charlier separated N_2, O_2, A, CO, CO_2, H_2S, COS and SO_2 on Porapak Q at 75°C and $-65°C$.[41] To condition the column 20 ppm SO_2 was added to the helium carrier gas. The two chambers of the thermistor detector employed in this work were alternately used as reference and sensing chamber for the low and higher temperature columns. The resolution obtained on a single chromatogram was in the following sequence: $N_2 + O_2 + A + CO$, CO_2, H_2S, COS, SO_2, N_2, O_2, A, CO.

With the aid of a column selector and three different columns, Di Lorenzo analysed trace amounts of O_2, CO, CH_4, CO_2, C_2H_4 and C_2H_6 in N_2 using a single injection of the sample.[42] The columns employed were as follows:

Column 1 (Porapak Q, length 1·5 m) for the separation of CH_4, CO_2, C_2H_4 and C_2H_6

Column 2 (Molecular sieve 5 Å, length 1 m) for the separation of O_2, N_2 and CO

Column 3 (Porapak Q, length 1 m) served to minimize the fluctuations in pressure and flow rate during the column change-over.

The method was developed as a rapid routine analysis, and a typical chromatogram is shown in Fig. 68.

The trace analysis of chemically reactive gases such as NO, NO_2, SO_2, H_2S and NH_3 is particularly difficult owing to their reactivity, especially with water vapour, and their strong adsorption on the metal and glass surfaces in the apparatus, which can make analysis impossible. Similarly, reaction can occur with the stationary phase, so that these compounds can only be analysed on certain columns. Morrison et al. determined traces of NO_2 in the range 5–150 ppm in N_2 and O_2 with methyl silicone oil on Fluoropak 80 and using an EC detector.[31] Traces of water were removed by means of a low temperature trap (dry ice–acetone), and the column was deactivated with small amounts of NO_2 before beginning the quantitative analyses. An EC detector having a pulsed voltage was used for the determination of NO_2 at concentrations below 5 ppm. The limit of detection attained for this method was 1 ppm NO_2.[43]

Separation of traces of NO in N_2 can be accomplished on silica gel which has been pretreated with NO_2.[44] The analytical limit is believed to be about 50 ppm NO.

Using a thermal conductivity detector and pretreated 5 Å molecular sieve, Dietz determined down to 12 ppm NO, 6 ppm CO, 25 ppm N_2O and 4 ppm CO_2.[45] To condition the column packing it was heated in vacuum for twenty hours at 300°C. Subsequently, helium was introduced at the same temperature, and then it was saturated with NO by cooling to room temperature in a

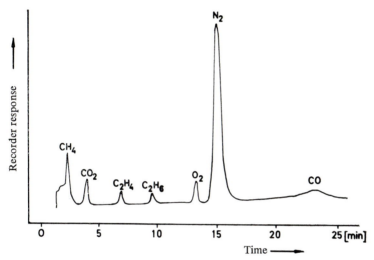

Fig. 68   Simultaneous determination of traces of inert gases and hydrocarbons in nitrogen (from A. Di Lorenzo, *J. Chromatog. Sci.*, **8**, 224 (1970)).

current of this gas. Excess NO was removed from the column by flushing with helium at 20°C. The actual analysis involved temperature programmed operation of the column.

Concentrations of NO_2 in N_2 and O_2 in the ppm range have been measured by Morrison *et al.*,[31] Greene and Pust,[46] and Trowell.[47] Wilhite and Hollis have shown that it is possible to obtain very good separation of various gases from the oxides of nitrogen.[40] Lawson and McAdie have described a method, in which an EC detector was used, of determining less than 10 ppm of the different oxides of nitrogen in air.[48] A detailed study was made of such problems as irreversible adsorption of NO and NO_2 on the column packing, and chemical interaction of water with these compounds, and also of air oxidation of NO on the column.

It is possible to determine small amounts of gaseous HCl in a mixture with H_2S and H_2O vapour if a Teflon detector block and tungsten filaments are used. The column is conditioned with gaseous HCl.[49] A chromatographic method of measuring the purity of ammonia, in which less than 3 ppm O_2, N_2, CO, CO_2, CH_4 and H_2O are determined, is described by Mindrup and Taylor.[50] Priestley *et al.* used an EC detector for the continuous determination of phosgene in the range 1 ppb to 2 ppm in air.[51] Figure 69 shows that this analysis can be carried out every six minutes.

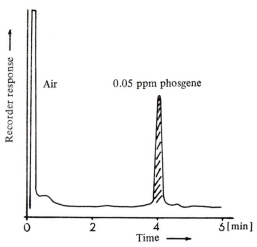

Fig. 69   Gas chromatogram of 0·05 ppm phosgene in air on a 2 m didecyl phthalate column at 50°C (from L. J. Priestley, F. E. Critchfield, N. H. Ketcham and J. D. Cavender, *Anal. Chem.*, **37**, 70 (1965)).

Small amounts of water vapour interfere with the analysis. On the other hand, HCl concentrations in the region of about 1% do not interfere. This detector is extremely sensitive to carbon tetrachloride, but it does not elute until after 30 minutes so that five phosgene determinations can be performed successively.

References
1. CONDON, R. D., SCHOLLY, P. R. and AVERILL, W., in *Gas chromatography 1960* (Ed. R. P. W. Scott), p. 30, Butterworths, London (1960).
2. ENGEL, R., *TIPS 36 GC*, Bodenseewerk Perkin Elmer, January 1968.
3. ZOCCHI, F., *J. Gas Chromatog.*, **6**, 251 (1968).
4. SCHWENK, U., HACHENBERG, H. and FÖRDERREUTHER, M., *Brennstoff-Chem.*, **42**, 295 (1961).
5. SABATIER, P. and SENDERENS, I. B., *Chem. Zentr.*, 974 (1902).
6. PORTER, K. and VOLMAN, D. H., *Anal. Chem.*, **34**, 748 (1962).
7. BEREZKIN, V. G., MYSAK, A. E. and POLAK, L. S., *Izv. Akad. Nauk SSSR, Ser. Khim.*, **10**, 1871 (1964).
8. GUDZINOWICZ, B. J. and SMITH, W. R., *Anal. Chem.*, **35**, 465 (1963).
9. HACHENBERG, H., unpublished work.
10. HOFFMANN, R. L. and EVANS, C. D., *J. Gas Chromatog.*, **4**, 198 (1966).
11. HACHENBERG, H. and GUTBERLET, J., *Brennstoff-Chem.*, **49**, 242 (1968).

12. DIMBAT, M., PORTER, P. E. and STROSS, F. H., *Anal. Chem.*, **28**, 290 (1956).
13. *Temperature-dependent resistors*, data sheet from the Valvo manual 1967, Valvo GmbH., 2 Hamburg 1.
14. LOVELOCK, J. E., *J. Chromatog.*, **1**, 35 (1958).
15. LIPSKY, S. R., LANDOWNE, R. A. and LOVELOCK, J. E., *Anal. Chem.*, **31**, 852 (1959).
16. WILLIS, V., *Nature*, **184**, 894 (1959).
17. LESSER, R., *Angew. Chem.*, **72**, 775 (1960).
18. BOTHE, H.-K. and LEONHARDT, J., *J. Chromatog.*, **19**, 1 (1965).
19. ELLIS, J. F. and FORREST, C. W., *Anal. Chim. Acta*, **24**, 329 (1961).
20. KARMEN, A., GIUFFRIDA, L. and BOWMAN, R. L., *J. Chromatog.*, **9**, 13 (1962).
21. BOYD, C. M. and MEYER, A. S., Report No. AEC ORNL 3619 Reprint file 309, 360 (1964).
22. BERRY, R., *Nature*, **188**, 579 (1960).
23. BOURKE, P. J., DAWSON, R. W. and DENTON, W. H., *J. Chromatog.*, **14**, 387 (1964).
24. GNAUCK, G., *Z. Anal. Chem.*, **189**, 124 (1962).
25. HARTMANN, C. H. and DIMICK, K. P., *J. Gas Chromatog.*, **4**, 163 (1966).
26. POY, F. and VERGA, R., URSS–ITALIA Symposium held in Tibilisi, 27–29 October 1970.
27. ANON., Varian Aerograph, manufacturer's publication 10/68/550 A-1007 S15 (1968).
28. LAMBERT, C., 'Gas chromatography and its application to gas analysis for the detection of traces', Symposium on *Graphite physics, diffusion and carbon transport* held at Harwell, 25 and 26th June 1963, published by the Dragon Project under Dragon Project Report 253.
29. LEIPNITZ, E. and STRUPPE, H. G., *Handbuch der Gaschromatographie*, Verlag Chemie GmbH., Weinheim/Bergstr. (1967), pp. 221–222.
30. BETHEA, R. M. and MEADOR, M. C., *J. Chromatog. Sci.*, **7**, 655 (1969).
31. MORRISON, M. E., RINKER, R. G. and CORCORAN, W. H., *Anal. Chem.*, **36**, 2256 (1964).
32. ISBELL, R. E., *Anal. Chem.*, **35**, 255 (1963).
33. PRIESTLEY, L. J., CRITCHFIELD, F. E., KETCHAM, N. H. and CAVENDER, J. D., *Anal. Chem.*, **37**, 70 (1965).
34. RUNGE, H., *Z. Anal. Chem.*, **189**, 111 (1962).
35. HUILLET, F. D. and URONE, P., *J. Gas Chromatog.*, **4**, 249 (1966).
36. CIEPLINSKI, E. W., Application No. GC-DS-003, Perkin Elmer, Corp. (1964).
37. JONES, C. N., *Anal. Chem.*, **39**, 1858 (1967).
38. ROBBINS, L. A., BETHEA, R. M. and WHEELOCK, T. D., *J. Chromatog.*, **13**, 361 (1964).
39. HOLLIS, O. L. and HAYES, W. V., in *Gas chromatography* 1966 (Ed. A. B. Littlewood), p. 57, The Institute of Petroleum, London (1967).
40. WILHITE, W. F. and HOLLIS, O. L., *J. Gas Chromatog.*, **6**, 84 (1968).
41. OBERMILLER, E. L. and CHARLIER, G. O., *J. Chromatog. Sci.*, **7**, 580 (1969).
42. DI LORENZO, A., *J. Chromatog. Sci.*, **8**, 224 (1970).
43. MORRISON, M. E. and CORCORAN, W. H., *Anal. Chem.*, **39**, 255 (1967).
44. SAKAIDA, R. R., RINKER, R. G., CUFFEL, R. F. and CORCORAN, W. H., *Anal. Chem.*, **33**, 32 (1961).
45. DIETZ, R. M., *Anal. Chem.*, **40**, 1576 (1968).

46. GREENE, S. A. and PUST, H., *Anal. Chem.*, **30**, 1039 (1958).
47. TROWELL, J. M., *Anal. Chem.*, **37**, 1152 (1965).
48. LAWSON, A. and McADIE, H. G., *J. Chromatog. Sci.*, **8**, 731 (1970).
49. OBERMILLER, E. L. and CHARLIER, G. O., *Anal. Chem.*, **39**, 396 (1967).
50. MINDRUP, R. F. Jr and TAYLOR, J. H., *J. Chromatog. Sci.*, **8**, 723 (1970).
51. PRIESTLEY, L. J. Jr, CRITCHFIELD, F. E., KETCHAM, N. H. and CAVENDER, J. D., *Anal. Chem.*, **37**, 70 (1965).

2.12. Organic Gases and Vapours

Trace analytical investigation of gases and materials in the gaseous state is a frequently occurring problem in gas chromatography. Thus, trace components in process streams or air separation plants require continual monitoring for safety reasons or because of corrosion. Analysis of air is becoming more and more important, whether in the atmosphere over cities or in factories and laboratories. In addition, all those gases should be considered which are used as starting materials for the manufacture of aliphatic intermediates.

Determination of hydrocarbons in air separation plants is of utmost importance for safety reasons. Their presence can become dangerous if their normally very low concentration in the air is increased in the fractionating unit. If such an installation is in a region where several refineries and petrochemical plants are concentrated in a small area then this danger is rather serious.

Instruments exist which use a flame ionization detector to indicate the total concentration of hydrocarbons in air.[1] They have limits of detection down to 1 ppb. However, it is not possible to differentiate between individual compounds since they do not incorporate a chromatographic column, and therefore only indicate the sum of the organic materials present. This kind of equipment therefore has only a limited value since in many cases it is desirable to know the nature of the individual hydrocarbons. The various analytical instruments suitable for determination of hydrocarbons in air separation plants have been summarized by Klein.[2] Figure 70 shows a chromatogram of traces of methane, ethane, ethylene, propane, acetylene and propylene at a level of 0·05 ppm in nitrogen.

It is worth noting that carbonyls should be removed from the hydrogen used in the FID by passing it through activated charcoal. Kuley determined the C_1 to C_4 hydrocarbons in air and liquid oxygen from air separation plants.[3] The limits of detection for the β-radiation detector used were reported to be 0·2–0·08 ppm in a 10 ml sample, and 0·07–0·004 ppm for the FID which was used simultaneously. The column packing was alkali-modified aluminium oxide coated with Carbowax. Use of untreated alumina or silica gel leads to errors due to polymerization and isomerization effects on temperature programming from 5 to 150°C.

Analyses for air separation plants represent only a small fraction of the problems associated with control of air purity. Thus, continuous monitoring of solvent vapours in the air in factories is also of the greatest importance for health reasons. While for a long time it has been possible exactly to measure excessive noise levels in decibels, until now there has only been the sense of smell as a means of detecting pollution due to vapours. Such noxious compounds can now also be measured by means of gas chromatography, and this provides valuable help in assigning MPL values (MPL = maximum permissible level). This involves detecting a large number of compounds such as various types of hydrocarbons, chlorinated hydrocarbons, alcohols, esters, ketones, acids, unsaturated aldehydes, etc. Whereas infrared and ultraviolet spectroscopy and many classical methods of chemical analysis detect only

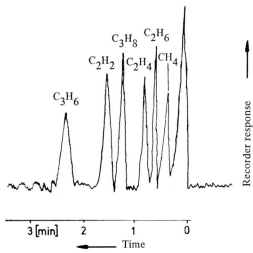

Fig. 70 Determination of trace amounts of hydrocarbons (from G. Klein, *Linde Berichte aus Technik und Wissenschaft*, **17**, 24 (1964)).

groups of compounds, the great advantage of gas chromatography lies in its ability to provide a complete analysis of such impurities. This may be achieved by simultaneously operating different specific detectors to identify the compounds concerned. In this connection, use of specific indicator tubes (Drägerwerk, Lübeck, West Germany), particularly at the column outlet, can be of considerable, and often indispensable assistance in the qualitative identification of the above types of compounds. A few of these materials with their MPL values are listed below.

Benzene	25 ppm	Carbon tetrachloride	10 ppm
Chloroacetaldehyde	1 ppm	Acetic anhydride	5 ppm
Crotonaldehyde	2 ppm	Phenol	5 ppm
Formaldehyde	5 ppm	Pyridine	5 ppm
Cresols	5 ppm	Nitrobenzene	1 ppm
Methyl chloride	50 ppm	Acrolein	0·1 ppm
Naphthalene	10 ppm	Ethyleneimine	0·5 ppm

Grupinski has employed infrared spectroscopic analysis in a 10 m long gas cell to determine 5×10^{-4} ml of ethyl acetate vapour in helium after separation by gas chromatography.[4] Separation of a large number of volatile organic compounds was examined on a series of columns such as Emulphor O, silicone grease and LAC-2-R-446 polyester, and the quantitative results obtained were compared with those determined by gas detector tubes.

If trace impurities are present in a very low concentration then they must be enriched. This can be done *e.g.* in heptane.[5] It is often necessary to employ temperature programming because a large number of different compounds may encompass a relatively wide range of boiling points as, for example, in the determination of methanol, isopropanol, toluene and the cresols.

May described the analysis of traces of various solvent vapours in air, and detected amounts down to 0·01 ppm with an FID.[6] A propylene glycol column was used at 100 and 50°C. The following list gives the concentrations in ppm for a 1 cm peak height for the various solvent vapours.

Solvent	1 cm peak height in ppm
Methyl formate	61
Acetone	35
Methyl acetate	56
Methanol	288
Ethanol	97
Isopropanol	57
Methyl ethyl ketone	41
Ethyl acetate	46
Tetrahydrofuran	50
1,1,1-Trichlorethane	84
Benzene	27
n-Propanol	95
Diethyl ketone	41
Dioxan	126
Isobutanol	64
Isobutyl acetate	45
Toluene	39
n-Butanol	77
Butyl acetate	58
Ethylene glycol	250
m-Xylene *p*-Xylene *o*-Xylene	67 (sum of the three peaks)

Compared with the concentrations of impurities occurring in factory air, those in the open air are considerably lower. Their detection therefore requires a very sensitive detector, or the use of a pre-concentration technique.

Cropper and Kaminsky concentrated the impurities in air in 10–15 minutes using a small tube packed with an adsorbent.[7] Subsequent gas chromatographic analysis showed the presence of toluene, benzyl chloride, benzal chloride, benzotrichloride and benzaldehyde, all of which originated from a chlorination plant.

Determination of vehicle exhaust gases in air is a specialized and very important aspect of gas chromatographic trace analysis, but which can only be briefly discussed here. Appropriate techniques have been developed, particularly by Altshuller and Clemons, an example being the determination of individual aromatic hydrocarbons in the range 0·05–1 ppm.[8] Eggertsen and Nelson have measured exhaust gases in city air after concentrating in a cold trap at liquid nitrogen temperature.[9] Jacobs has reported a rapid gas chromatographic determination of C_1–C_{10} hydrocarbons in exhaust gases, the analysis involving a capillary column, temperature programmed from $-55°C$ to $140°C$.[10] More than 85 C_1–C_{10} paraffins, olefins and aromatics at concentrations below 1 vol ppm were determined in 13 minutes by this method.

Owing to the large number of publications, it is again only possible to mention briefly the numerous successful applications of gas chromatography in the determination of hydrocarbons and other constituents in cigarette smoke. In these investigations, combination of mass spectrometry with gas chromatography has been especially valuable, and it is the only method which has enabled the nature of these particular trace compounds to be elucidated.[11] Using this method, Grob and Voellmin have been able to identify 133 components in cigarette smoke.[12] Even before the introduction of flame ionization detectors, trace components in air could be determined at levels below 1 ppm by use of pre-concentration techniques. After separating and concentrating the traces of hydrocarbons by gas chromatography, Heaton and Wentworth burned them to produce CO_2 and H_2O.[13] As a detector they used an infrared instrument sensitized to be specific for CO_2. Quantitatively reproducible combustion of such trace materials is not easily achieved, and variations in the CO_2 concentration in the sample and air must be taken into account. Removal of water formed during combustion also presents problems.

Lawrey and Cerato analysed 1–6000 vol ppm methane in air after pre-concentration of the sample.[14] The method was reported to have a precision of ± 2 ppm at a 30 ppm level. An amount of 0·1 ppm of acetylene may be determined by concentrating the sample on silica gel followed by chromatography.[15]

Determination of hydrocarbons in air at the ppb level has been reported by Bellar et al.[16,17] To identify atmospheric impurities, Williams has used a special technique based on microreactions (see 1.351), after collection of the air sample at $-80°C$.[18] In this procedure, the sample had already undergone a preliminary separation since two alternative drying agents were used. Treatment with K_2CO_3 let most of the components through, while $Mg(ClO_4)_2$ retained certain types of compound. An additional column located between the collecting and chromatographic columns provided a further means of classification before the actual gas chromatographic separation. When this column was packed with an inert support coated with concentrated sulphuric acid, it retained unsaturated hydrocarbons, sulphides and a number of other organic compounds by reacting with them, whilst not holding back saturated hydrocarbons. It was possible subsequently to separate these into n-paraffins, isoparaffins and cycloalkanes with 5 Å molecular sieve. These groups of hydrocarbons were then chromatographed on two different selective stationary phases, didecyl phthalate and tritolyl phosphate, using two different specific detectors, FID and ECD.

The same technique was described in detail by Farrington et al.[19] One of the main problems in the pre-concentration procedures is to remove water vapour without losing organic material. Various drying agents such as 4 Å molecular sieve, $CaSO_4$, $Mg(ClO_4)_2$, $Ba(ClO_4)_2$ and K_2CO_3 were tested in this connection with the vapours of a variety of alcohols, ethers, ketones, aldehydes, etc. It was found that only hydrocarbons diffuse through all of the drying agents, while the rest of the compounds are more or less strongly retained, so that an incorrect analytical result can even originate in sampling.

It is advisable to employ electron capture detectors if polynuclear hydrocarbons are to be determined in mixtures containing other hydrocarbons. It is then possible to measure, for example, less than 10^{-9} g of benzpyrene in air, the separation being carried out on silicone grease at 215°C.[20]

Knipschild and Pagnier have described the analysis of naphthalene and 1- and 2-methylnaphthalene in coke-oven and town gas.[21] They concentrated these components at -50°C prior to gas chromatographic analysis. Pre-concentration of polynuclear aromatics present in air can also be carried out on glass fibre filters, followed by extraction with various solvents. This technique may be used to determine pyrene, chrysene, benzanthracene and other polynuclear hydrocarbons by temperature programmed gas chromatography, although a second column must be used to compensate for bleeding of the stationary phase.[22] Naphthalene, anthracene, pyrene and a large number of other polynuclear hydrocarbons may be identified by a combination of gas and thin layer chromatography.[23] The C_2–C_5 hydrocarbons can be determined in air by using a dual FID, and further separation after a didecyl phthalate column by means of a tube packed with mercuric perchlorate, which only lets the paraffins through.[24]

Camphene can be determined in air by concentrating in carbon tetrachloride and using xylene as an internal standard for the quantitative analysis.[25] Himmelreich has reported an analytical procedure for the determination of non-aromatic hydrocarbons, benzene, toluene, ethylbenzene and the xylenes in petrol–air mixtures.[26] Brenner and Ettre used a pre-concentration technique in the determination of acetylene in either air or oxygen.[15] The limit of detection with a thermal conductivity detector was 0·1 ppm for a sample size of 10 litres.

Another very interesting example, in the field of food science, is the determination of traces of ethylene in air. Ethylene is used to accelerate the ripening of apples, and, moreover, some fruits produce ethylene themselves. Because of its high sensitivity, gas chromatography provides an excellent means of investigating these phenomena.[27] In similar studies, Chon-Ton Phan showed that traces of ethylene are generated by certain flowers and

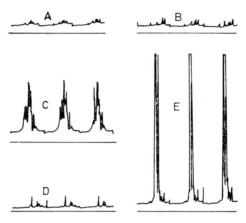

Fig. 71   Chromatograms obtained with a reversion gas chromatograph, (*A*) forest air, (*B*) air in a chemical factory, (*C*) compressed air from a normal cylinder, (*D*) synthetic air, (*E*) laboratory air (from R. Kaiser, *Chromatographia*, **1**, 206 (1968)).

some types of fruit.[28] These few examples of the application of gas chromatography to the analysis of air would be incomplete without reference to reversion gas chromatography, which is the most sensitive method available at present (*see* 1.343). It enables trace components to be detected at the ppb level. Figure 71 shows the application of this method to the analysis of samples of air taken from a forest, factory, physics laboratory, and compressed air cylinder and also a sample of synthetic air.[29]

In addition to the trace analysis of air, which could only be described briefly owing to the great number of publications, determination of organic gases and vapours in other gases is of considerable importance.

To determine ppb amounts of hydrocarbons in cylinder gases such as helium, nitrogen and hydrogen, Feldstein and Balestrieri recommended a pre-concentration by cooling 150 ml samples to liquid nitrogen temperature.[30] This method of concentration gives a 95–100% yield. The subsequent gas chromatographic analysis was performed with an FID.

If gases containing chlorinated hydrocarbons are to be analysed then an ECD is preferred to any other type of detector for sensitivity of detection. Figure 72 shows the determination of traces of carbon tetrachloride and chloroform in hydrogen.[31]

Mirzayanov et al. determined traces of C_1–C_4 hydrocarbons in oxides of nitrogen with an Al_2O_3 column coated with sodium hydroxide and used at 120°C in association with an FID.[32] Catalytic reduction of the oxides of nitrogen to nitrogen and water prevented interference with the detector signal.

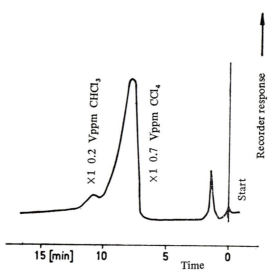

Fig. 72   Determination of 0·7 vol ppm carbon tetrachloride and 0·2 vol ppm chloroform in hydrogen using a 3 m tricresyl phosphate column at 50°C and an ECD.

The method was reported to have an accuracy of $\pm 3·5\%$ at the level of 30 vol ppm. The C_1–C_3 hydrocarbons have been analysed in electrolytic hydrogen by Brenner and Ettre on a 2 m silica gel column after pre-concentration in the gas sampling system.[15] The C_4 and C_5 hydrocarbons were separated on a 4 m dimethylsulpholane column.

Freund et al. have improved the gas chromatographic analysis of traces of hydrocarbons in various technologically important products such as carbon dioxide, natural gas and cracking gas.[33] Using Apiezon L as the stationary phase at 100°C, 10 ml samples of crude carbon dioxide were analysed for hydrocarbons. Since an FID was used, and it did not respond to CO_2, the limits of detection for the hydrocarbons were extremely good. In addition to the gas chromatographic analysis, an olfactory test was carried out. This was done by expanding the gas to produce lumps of solid carbon dioxide. After 20–30 minutes exposure to the sample their surface was wiped with cotton wool. The trace impurities concentrated in this way may easily be detected down to 0·1 ppm by their characteristic odour. It was established that the unpleasant odour usually associated with dry ice is due to hydrocarbons having 5 or more carbon atoms. Thus, *e.g.* 20 ppm C_5, 7·3 ppm C_6, 9·2 ppm C_7, 5·5 ppm C_8 and 0·9 ppm C_9 hydrocarbons produce a strong odour in gaseous carbon dioxide, and even 0·9 ppm of C_5 hydrocarbons is distinctly noticeable.

Determination of high molecular weight hydrocarbons in natural and cracking gas is also of considerable importance in the industrial use of these materials. Freund specified the following gas chromatographic conditions for the analysis of natural gas:

Analysis of C_1–C_4 hydrocarbons: column: 1 m alumina
Analysis of C_5 and higher hydrocarbons: column: 0·8 m Apiezon L at 70°C
Analysis of aromatics: column: 2 m β,β'-oxydipropionitrile.

The analysis of methane from a cracking plant on an alumina column at three different temperatures is shown in Fig. 73.

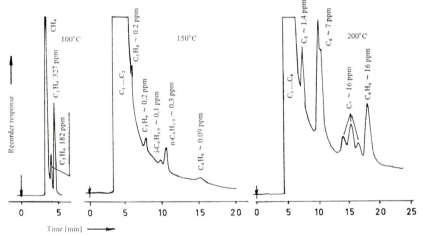

Fig. 73   Analysis of methane from a cracking plant (from M. Freund *et al.*, *Erdöl Kohle*, **17**, 996 (1964)).

Zocchi has pointed out that it is difficult to determine traces of hydrocarbons in methane in the ppm region since the methane peak can overlap the components being analysed.[34] Thus, it was found experimentally that, in a 170 ml sample of methane containing 3·2 ppm ethane, 0·75 ppm ethylene, 0·5 ppm propane and 3 ppb propylene, only the propane can be detected. A preliminary separation of these trace components from the main constituent at liquid nitrogen temperature gave a 10^4 fold reduction in the methane concentration. This enabled the C_2–C_5 hydrocarbons to be determined down to the ppb level.

While determination of traces of C_4 in C_1–C_3 hydrocarbons does not involve any sampling problems, the reverse case, *i.e.* determination of C_1–C_3 in liquid C_4 hydrocarbons, is more difficult since sampling must then be carried out in the liquid state (*see* 1.331). Greater difficulty also occurs with regard to resolution when determining the purity of C_4 hydrocarbons, since considerably more trace components are involved than with C_2 and C_3 hydrocarbons. Examples of this are given in the following section.

References
1. ANDREATCH, A. J. and FEINLAND, R., *Anal. Chem.*, **32**, 1021 (1960).
2. KLEIN, G., *Linde Berichte aus Technik und Wissenschaft*, **17**, 24 (1964).
3. KULEY, C. J., *Anal. Chem.*, **35**, 1472 (1963).
4. GRUPINSKI, L., *Wasser, Luft, Betrieb*, 77 (1966).
5. BERNET, J., *Wasser, Luft, Betrieb*, 396 (1966).
6. MAY, J., *Staub*, **25**, 153 (1965).
7. CROPPER, F. R. and KAMINSKY, S., *Anal. Chem.*, **35**, 735 (1963).
8. ALTSHULLER, A. P. and CLEMONS, C. A., *Anal. Chem.*, **34**, 466 (1962).
9. EGGERSTEN, F. T. and NELSON, F. M., *Anal. Chem.*, **30**, 1040 (1958).
10. JACOBS, E. S., *Anal. Chem.*, **38**, 43 (1966).
11. GELPI, E. and ORÓ, J., *J. Chromatog. Sci.*, **8**, 210 (1970).
12. GROB, K. and VOELLMIN, J. A., *J. Chromatog. Sci.*, **8**, 218 (1970).
13. HEATON, W. B. and WENTWORTH, J. T., *Anal. Chem.*, **31**, 349 (1959).
14. LAWREY, D. M. G. and CERATO, C. C., *Anal. Chem.*, **31**, 1011 (1959).
15. BRENNER, N. and ETTRE, L. S., *Anal. Chem.*, **31**, 1815 (1959).
16. BELLAR, T. A., SIGSBY, J. E., CLEMONS, C. A. and ALTSHULLER, A. P., *Anal. Chem.*, **34**, 763 (1962).
17. BELLAR, T. A., BROWN, M. F. and SIGSBY, J. E., *Anal. Chem.*, **35**, 1924 (1963).
18. WILLIAMS, I. H., *Anal. Chem.*, **37**, 1723 (1965).
19. FARRINGTON, P. S., PECSOK, R. L., MEEKER, R. L. and OLSON, T. J., *Anal. Chem.*, **31**, 1512 (1959).
20. DUCAN, R. M., *Am. Ind. Hyg. Assoc. J.*, **30**, 624 (1969).
21. KNIPSCHILD, J. and PAGNIER, G., *Brennstoff-Chem.*, **44**, 8 (1963).
22. DeMAIO, L. and CORN, M., *Anal. Chem.*, **38**, 131 (1966).
23. CHATOT, G., JEQUIER, W., JAY, M., FONTANGES, R. and OBATON, P., *J. Chromatog.*, **45**, 415 (1969).
24. GORDON, R. J., MAYRSOHN, H. and INGELS, R. M., *Environmental Sci. and Technol.*, **2**, 1117 (1968).
25. LIEBMANN, R. and HENNINGS, P., *Chem. Tech. (Berlin)*, **19**, 44 (1967).
26. HIMMELREICH, J. J., *Erdöl Kohle*, **23**, 366 (1970).
27. MEIGH, D. F., *J. Sci. Food Agr.*, **11**, 381 (1960).
28. PHAN, CHON-TON, *Phytochemistry*, **4**, 353 (1965).
29. KAISER, R., *Chromatographia*, **1**, 199 (1968).
30. FELDSTEIN, M. and BALESTRIERI, S., *J. Air Poll. Control Ass.*, **15**, 177 (1965).
31. HACHENBERG, H. and GUTBERLET, J., *Brennstoff-Chem.*, **49**, 242 (1968).
32. MIRZAYANOV, V. S., BEREZKIN, V. G. and NIKOL'SKII, V. A., *Zh. Analit. Khim.*, **21**, 1239 (1966).
33. FREUND, M., SZEPESY, L. and SIMON, J., *Erdöl Kohle*, **17**, 995 (1964).
34. ZOCCHI, F., *J. Gas Chromatog.*, **6**, 100 (1968).

2.13. Higher Acetylenes and 1,2-Diolefins in Gaseous Mixtures

The thermal cracking of hydrocarbons produces C_2 to C_4 unsaturated hydrocarbons containing acetylenic and 1,2-diolefinic constituents as impurities. From the point of view of safety and operating life of production plants, it is vitally important to determine these components. Moreover, regarding the subsequent purity required for the olefinic monomers, it is advantageous to have already identified the impurities in the crude gas from the cracker. This involves determination of small amounts of higher acetylenes, 1,3-butadiene and cyclopentadiene in gaseous mixtures which can contain up to 30 components.

This task can only be carried out by gas chromatography since the chemical methods available, *e.g.* the formation of acetylides from heavy metal ions and acetylenic hydrocarbons, only permit detection of groups of compounds. Moreover, non-terminal acetylenes are not detected by this method.

Scoggins and Price have described a method of group analysis which detects C_4–C_5 acetylenes in hydrocarbons by hydration to form carbonyl compounds followed by ultraviolet spectroscopic determination of the 2,4-dinitrophenylhydrazones.[1] This method also detects acetylenes with non-terminal triple bonds.

As a result of the above difficulties, gas chromatographs were first employed for this analysis about 15 years ago, and existing classical chemical and spectroscopic techniques very rapidly superseded. Only by gas chromatography can the above compounds selectively be determined in complex mixtures, and the conditions must be chosen so that each component can be accurately determined without masking by other hydrocarbons. As is shown

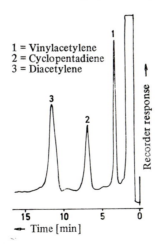

1 = Vinylacetylene
2 = Cyclopentadiene
3 = Diacetylene

Fig. 74 Determination of vinylacetylene, cyclopentadiene and diacetylene in cracking gas (from E. Schneck, *Brennstoff-Chem.*, **44**, 359 (1963)).

below, this is not a simple task, and requires the use of several stationary phases.

As shown in Fig. 74, cyclopentadiene, diacetylene and vinylacetylene may be determined in cracking gas in 15 minutes on a 4 m β,β'-oxydipropionitrile column.[2]

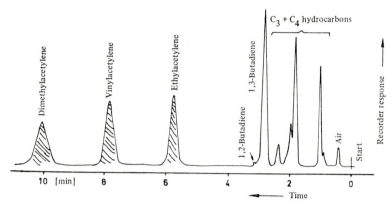

Fig. 75 Determination of higher acetylenes in C_4 hydrocarbons (3 m β,β'-oxydipropionitrile) (from S. A. Pollard, *Anal. Chem.*, **36**, 999 (1964)).

Using the above stationary phase and also a sulpholane column, Pollard has measured vinylacetylene in C_4 fractions.[3] Columns 3 m long were employed, and gave good resolution for vinylacetylene, enabling 5 vol ppm to be detected. The reproducibility was reported to be within 6 ppm in the region of 36 ppm, and within 48 ppm at the 1000 ppm level. Figure 75 shows

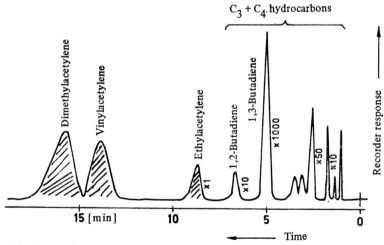

Fig. 76 Determination of higher acetylenes in C_4 hydrocarbons with a 3·6 m dimethylsulpholane column.

the determination of vinylacetylene, ethylacetylene and dimethylacetylene in
C_4 hydrocarbons on a β,β'-oxydipropionitrile column, and Fig. 76 shows the
same analysis on a dimethylsulpholane column.

If a considerably longer analysis time is acceptable, then it is possible to
determine diacetylene, vinylacetylene and methylacetylene on silica gel.[4] The
stability of this column packing is highly suitable if an FID is employed.

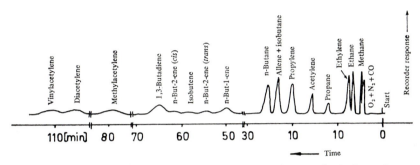

Fig. 77 Determination of methylacetylene, diacetylene and vinylacetylene on
silica gel.

Figure 77 shows the separation of these compounds on a 9·5 m long silica
gel column at 21°C. For this analysis the carrier gas must contain water
vapour which largely eliminates the memory and tailing effects usually
observed with silica gel. This is achieved by passing the carrier gas over 40%
sulphuric acid.

All of the C_4 acetylenes, as well as allene and methylacetylene, can also be
analysed on a perhydrophenanthrene column (*see* 2.51). Figure 78 shows that
acetonyl acetone is also a very good stationary phase for the analysis of

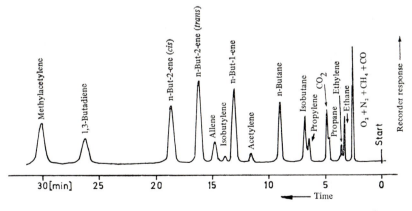

Fig. 78 Determination of allene and methylacetylene on a 22 m acetonyl acetone
column.

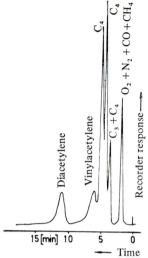

Fig. 79   Determination of diacetylene and vinylacetylene on a 3 m dioctyl phthalate column at 70°C.

allene and methylacetylene in C_4 hydrocarbons. In the absence of C_5 hydrocarbons, it is also possible to determine diacetylene and vinylacetylene on a dioctyl phthalate column (Fig. 79).

Determination of the higher acetylenes in other hydrocarbons has been discussed in detail by Scharfe.[5]

Figure 80 shows a chromatogram for the analysis of allene and methylacetylene, vinylacetylene and ethylacetylene, while the determination of dimethylacetylene is illustrated in Fig. 81.

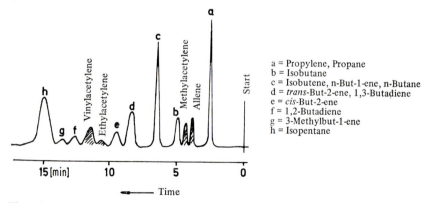

a = Propylene, Propane
b = Isobutane
c = Isobutene, n-But-1-ene, n-Butane
d = trans-But-2-ene, 1,3-Butadiene
e = cis-But-2-ene
f = 1,2-Butadiene
g = 3-Methylbut-1-ene
h = Isopentane

Fig. 80   Determination of allene and methylacetylene, ethylacetylene and vinylacetylene in C_3–C_5 hydrocarbons on a 6 m column containing a mixture of polar and non-polar liquids as stationary phase (from G. Scharfe, *Erdöl Kohle*, **12,** 723 (1959)).

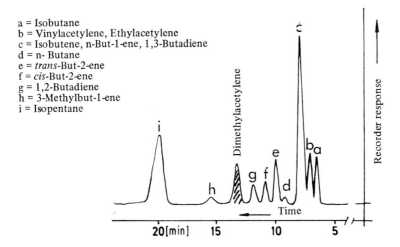

a = Isobutane
b = Vinylacetylene, Ethylacetylene
c = Isobutene, n-But-1-ene, 1,3-Butadiene
d = n- Butane
e = *trans*-But-2-ene
f = *cis*-But-2-ene
g = 1,2-Butadiene
h = 3-Methylbut-1-ene
i = Isopentane

Fig. 81 Determination of dimethylacetylene in C_4 and C_5 hydrocarbons on a 6 m column containing 9 parts of tetraisobutene to 1 part of Mepasin as stationary phase (from G. Scharfe, *Erdöl Kohle*, **12**, 723 (1959)).

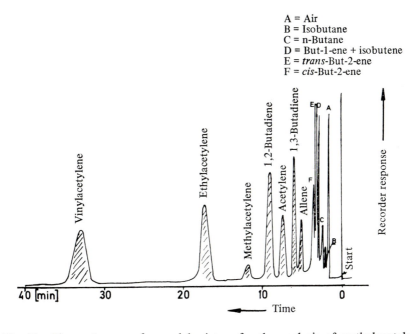

A = Air
B = Isobutane
C = n-Butane
D = But-1-ene + isobutene
E = *trans*-But-2-ene
F = *cis*-But-2-ene

Fig. 82 Chromatogram of a model mixture for the analysis of methylacetylene allene, ethylacetylene and vinylacetylene in C_2–C_4 hydrocarbons (from S. Rennhak *et al.*, *Chem. Tech.* (*Berlin*), **17**, 690 (1965)).

Rennhak *et al.* have also examined the use of several highly selective liquid phases for the analysis of traces of higher acetylenes in gaseous mixtures.[6] They found that dimethyl sulphoxide shows the best selectivity both for the separation of acetylenes from mono- and diolefins, and 1,2-dienes from mono-olefins. It was also found that only the dimethyl derivative has this high selectivity, since it is not observed with dipropyl sulphoxide. Figure 82 shows the chromatogram obtained with a model mixture of acetylenes in other hydrocarbons on a 6 m column of 25% dimethyl sulphoxide on Sterchamol with a thermal conductivity detector. The high selectivity of this stationary phase for acetylenes and cumulated 1,2-dienes is evident from the fact that acetylene elutes after 1,3-butadiene, and allene after *cis*-butene-2. However, the use of dimethyl sulphoxide for trace analysis is unfortunately limited owing to its relatively low boiling point of 189°C and relatively high vapour pressure (0·6 mm of mercury at 25°C). This means that when ionization detectors are used the column can only be operated at low temperatures. The usual technique for overcoming this difficulty is to install a cold trap between the column and detector. The temperature of such traps is generally between −80 and −130°C. This precludes the use of the technique in the present case since part of the C_4 trace components would be condensed out.

It is therefore preferable to employ involatile selective phases for the analysis of trace amounts of these hydrocarbons. The work of McKinnis has demonstrated that compounds which contain donor centres capable of forming hydrogen bonds with acetylene, in particular the $P=O$, $P—O—P$, $S=O$, $C—F$ groups, and also electron donating substituents such as $—N(CH_3)_2$, $—P(OCH_3)_2$, $—NC_4H_8$, $—CH_3$, $—C_2H_5$ and $—CH_2OCH_3$ are selective solvents for acetylenes.[7]

On the basis of these findings, Rennhak *et al.* synthesized two liquid phases, tripyrrolidinylphosphine oxide (vapour pressure: 5×10^{-3} mm of mercury at 85–100°C) and β-ethoxyvinylphosphonium-N,N'-dimethyldiamide (vapour pressure: 2 mm of mercury at 84–87°C).[6]

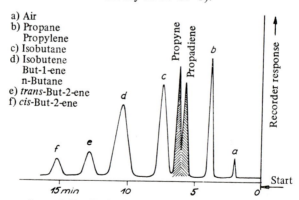

Fig. 83   Separation of C_3–C_4 hydrocarbons on an 8 m silicone oil column at 20°C (from G. Heuschkel *et al.*, *Erdöl Kohle*, **13**, 98 (1960)).

Heuschkel *et al.* determined propyne and propadiene, each with a limit of detection of 5 ppm in mixtures of propane and propylene containing up to 0·5% of C_4 hydrocarbons.[8] Since direct chromatography on silicone oil and diethylformamide columns did not give satisfactory results, the gaseous mixture was separated on a precolumn of silicone oil and the required fraction trapped out with liquid nitrogen and then resolved on the diethylformamide

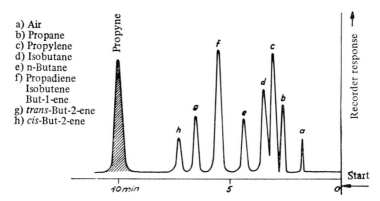

a) Air
b) Propane
c) Propylene
d) Isobutane
e) n-Butane
f) Propadiene
  Isobutene
  But-1-ene
g) *trans*-But-2-ene
h) *cis*-But-2-ene

Fig. 84   Separation of C_3–C_4 hydrocarbons on a 5 m diethylformamide column at 20°C (from G. Heuschkel *et al., Erdöl Kohle*, **13**, 98 (1960)).

column. Figure 83 shows the chromatogram of the eluate from the precolumn and Fig. 84 that for the main column, while the final chromatogram resulting from this two-stage procedure is presented in Fig. 85.

Kaufman and Zlatkis have determined traces of vinylacetylene and other C_4 hydrocarbons in 1,3-butadiene in less than 20 minutes using a 9·14 m column having 5% isosafrole as the stationary phase.[9]

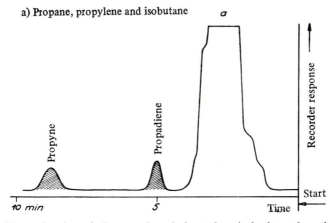

a) Propane, propylene and isobutane

Fig. 85   Determination of allene and methylacetylene in hydrocarbons by the two-stage method (from G. Heuschkel *et al., Erdöl Kohle*, **13**, 98 (1960)).

References
1. SCOGGINS, M. W. and PRICE, H. A., *Anal. Chem.,* **35,** 48 (1963).
2. SCHNECK, E., *Brennstoff-Chem.,* **44,** 354 (1963).
3. POLLARD, S. A., *Anal. Chem.,* **36,** 999 (1964).
4. HORN, O., SCHWENK, U. and HACHENBERG, H., *Brennstoff-Chem.,* **39,** 336 (1958).
5. SCHARFE, G., *Erdöl Kohle,* **12,** 723 (1959).
6. RENNHAK, S., DÖRING, C. E., SCHMID, G., SCHNELLER, D., STÜRTZ, H. and WERNER, E., *Chem. Tech. (Berlin),* **17,** 688 (1965).
7. McKINNIS, A. C., *Ind. Eng. Chem.,* **47,** 850 (1955).
8. HEUSCHKEL, G., WOLNY, J. and SKOCZOWSKI, S., *Erdöl Kohle,* **13,** 98 (1960).
9. KAUFMAN, H. R. and ZLATKIS, A., *Chem. & Ind. (London),* 1001 (1958).

2.14. Impurities in Acetylene

In addition to the old-established carbide process, manufacture of acetylene is carried out to an increasing extent by thermal cracking of hydrocarbons.[1] Different impurities have to be considered depending on the method of manufacture, and associated purification stages.

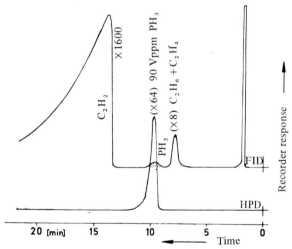

Fig. 86   Simultaneous chromatograms for the determination of 90 vol ppm PH_3 in C_2H_2 on a 2 m Porapak T column at 40°C with an FID and an HPD.

Purified gas from the carbide process contains, in addition to other materials, small amounts of hydrogen sulphide and phosphine. These two components produce the usual smell of acetylene and they are poisonous, particularly the phosphine. Selective determination of hydrogen sulphide may be performed with a flame photometric detector (*see* 2.4 and 2.51), whilst that of phosphine is carried out with the halogen–phosphorus detector (HPD). Figure 86 shows this method of analysis of PH_3 in acetylene, and that an FID can be used simultaneously to determine other impurities. This analysis can also be carried out continuously with a thermal conductivity detector by automatically injecting 10 ml samples onto a silica gel column.[3] In addition to the trace components mentioned, it must also be remembered that higher acetylenes such as diacetylene, vinylacetylene and ethylacetylene are present in C_2H_2 produced by the thermal cracking of hydrocarbons. Determination of these components is important, for example in cylinders containing acetylene where the danger exists, when the majority of the acetylene has been used, that these by-products may be concentrated. Figure 87 shows the analysis of a test mixture of 0·5 vol ppm of vinylacetylene and 8 vol ppm of ethylacetylene in acetylene.[4]

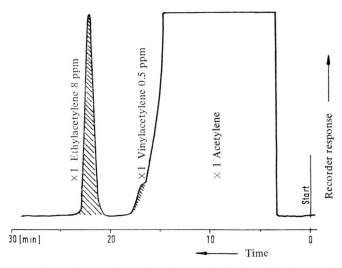

Fig. 87   Detection of traces of vinylacetylene and ethylacetylene in acetylene on an 8 m perhydrophenanthrene column at 22°C with an FID.

Tabuteau has developed an interesting method based on an infrared detector to determine the impurities mentioned above.[5] By this method, 50 ppm methylacetylene, 20 ppm monovinylacetylene and 10 ppm diacetylene can be determined.

Traces of oxygen, nitrogen, carbon monoxide and hydrogen may be easily analysed in the normal way on molecular sieve columns, and they are therefore not mentioned separately here.[6]

References
1. WINNACKER, K. and KÜCHLER, L., *Chemische Technologie,* Karl Hauser Verlag, Munich 1959, Volume 3, pp. 654–667.
2. HACHENBERG, H. and GUTBERLET, J., *Brennstoff-Chem.,* **49,** 242 (1968).
3. STEINDORF, W., JUST, E. and ARDELT, H. W., *Z. Chem.,* **5,** 388 (1965).

4. HACHENBERG, H., unpublished work.
5. TABUTEAU, J., *Ind. Chem. Belge Suppl.,* **1,** 135 (1959).
6. MILLER, S. A., *Acetylenes,* Ernest Benn Limited, London, 1965, pp. 347–349.

2.15. Head Space Analysis

Head space analysis is the analysis of the vapour phase in equilibrium with a liquid phase in a closed system. It is therefore a particular form of trace analysis of gases, and in principle does not represent a new analytical method, since determination of compounds in their vapour state above liquids and solids may be carried out by the appropriate technique, *e.g.* mass spectrometry and infrared and ultraviolet spectroscopy. However, these methods have the disadvantages of a sensitivity which is frequently too low and the fact that when vapours of several compounds are present an inextricable mass of information is obtained, since only the sum of the components present can be derived from the results.

A new aspect of head space analysis is its use in combination with gas chromatography. This enables the high resolution combined with great sensitivity of gas chromatographic ionization detectors to be utilized. Thus, head space analysis has again become of considerable importance, especially when trace components have to be analysed in samples which cannot normally be handled with a syringe, because of decomposition when vaporized or the formation of dissociation products not originally present in the sample.

Examples of applications of head space analysis are to solvent residues in chemicals and packaging materials, aromatics in tobacco and fruits and volatile components in medicines or buffer solutions, none of which can be analysed directly by gas chromatography, but only after concentration procedures have been carried out.

The first applications of head space analysis were in the foodstuffs and packaging industries. They made it possible to identify the flavours and aromas which could otherwise be characterized only by their taste or smell. Gas chromatographic head space analysis thus acts as a 'gas chromatographic nose' which in many cases can visibly indicate the odour in the form of peaks, and also the number of components which may be causing this odour. The differences in specificity between the human nose and the gas chromatographic detector should, of course, not be overlooked. MacKay *et al.* have examined the volatile components in the head gases over peppermint oil, bananas, roasted coffee, brandy, whisky, etc. and have been able in this way to establish the quality and genuineness of the aromas without having to carry out precise qualitative and quantitative analyses.[1] Figure 88 shows this kind of comparison between crushed fresh banana (chromatogram *A*) and an artificial banana flavouring (chromatogram *B*).

Brown *et al.* and Romani and Ku have respectively analysed the volatiles from apples and pears.[2,3] Drawert *et al.* have compared the head space technique with the usual liquid–liquid extraction with pentane–methylene chloride (2:1) for the gas chromatographic analysis of the aroma from apples.[4] They obtained different results from the two methods and concluded that head space analysis is very good for characterizing the composition of the substances producing the aromas, but it is unsuitable for their quantitative analysis.

Weurman has investigated the enzymatic formation of volatile components in ripe and unripe raspberries.[5] Other work of this kind has been concerned with the determination of NO, N_2O, N_2, O_2, CO_2, CO and H_2 in packaged foods, and with the quality control of fresh, frozen and packed vegetables.[6,7]

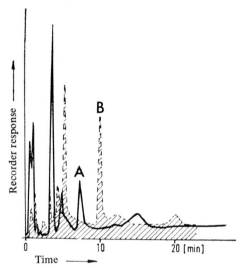

Fig. 88   Comparison of the head space chromatograms from crushed banana (*A*) and banana flavouring (*B*).

Miethke has analysed various kinds of alcoholic beverages such as cherry brandy, brandy, perry, cognac, egg flip, etc. with respect to their amyl alcohol content, appelation and adulterants.[8]

The head space method was used by Jentzsch *et al.* to distinguish camomile tea from peppermint tea.[9] As a result of its sensitivity, the technique can also rapidly provide information in other applications, for example, to distinguish used lubricating oil from the unused material.[10]

There is considerable interest in this analytical technique for the examination of plastics and polymer dispersions. Thus, for example, ethane, propane, n-butane and butene-1 can be detected as products of a radical degradation of polyethylene.[11] Since this is used as a packaging material for milk, these compounds can be a source of part of the plastic flavour in the milk.

Hachenberg has given examples of the application of head space analysis in the characterization of solid polymers and polymer dispersions.[12] He found that this technique can be more sensitive for the detection of residual monomers than the methods normally used, which require the polymer to be dissolved and reprecipitated. Figure 89 illustrates the conventional method of determining styrene in polystyrene by dissolving in CH_2Cl_2 and precipitation with methanol, followed by normal gas chromatographic sampling with a syringe.

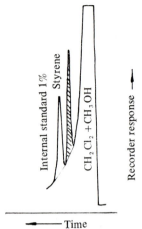

Fig. 89   Determination of styrene in polystyrene by dissolving in CH_2Cl_2 and precipitating with CH_3OH.

Except for the styrene peak, the only other identifiable peak is that due to the internal standard since the solvent and precipitant are eluted together. The analytical sample contains such a high proportion of these two components that other impurities cannot be determined because the signal due to CH_2Cl_2 and CH_3OH obscures the whole chromatogram. Also, this procedure

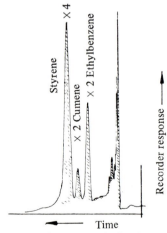

Fig. 90   Head space analysis of polystyrene.

has a low sensitivity for the detection of styrene. On the other hand, head space analysis provides considerably more information, as shown by Fig. 90, since other compounds are determined at the same time. Moreover, analysis can be carried out more rapidly since time-consuming preparation of the sample (dissolution and precipitation) is avoided.

Other analyses of this type *e.g.* of α-methylstyrene and polymethylstyrene, and also of a number of other polymers, copolymers and polymer dispersions represent many more possible applications of the technique, such as the determination of residual petroleum ether in polymers (Fig. 91).

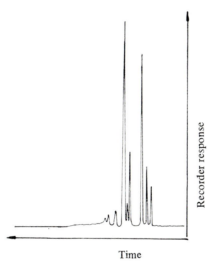

Time

Fig. 91 Determination of residual petroleum ether in polymers by gas chromatographic head space analysis.

Gas chromatographic head space analysis also provides a simple and rapid method for testing the chemical stability of polymers. Thus, for example, the resistance to water vapour, gaseous HCl, etc. may be studied at various temperatures in the actual sample container by analysing its vapours directly. Figure 92 shows the head space analysis of a polyacetal, and Fig. 93 the analysis of the same material after exposure to gaseous HCl for about half and hour.

No matter how convincing the above examples of this analytical technique may be, the considerable difficulties sometimes involved in both the qualitative and quantitative analyses should not be overlooked. It is extremely difficult, especially for commercial samples, to achieve a complete qualitative analysis by the head space technique, because the components in the vapour phase over the sample are only present at the ppm level, or frequently even at the ppb level. In many cases, it is thus not possible to specify the nature of the individual peaks. For liquid samples, therefore, Basette *et al.* carried out a

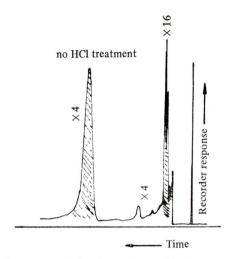

Fig. 92   Head space analysis of a polyacetal before exposure to HCl.

qualitative identification of groups of compounds from their reactions with specific reagents.[13] In this way, they were able to remove from the head space *e.g.* ketones by means of acidified hydroxylamine solution, sulphur compounds with mercuric chloride, and esters with alkaline hydroxylamine solution. Palo has described the same method of identification for determination of functional groups in the volatile constituents of dairy products.[14]

It is preferable to carry out the identification using the g.c.–m.s. combination, but only very sensitive mass spectrometers can be considered for this

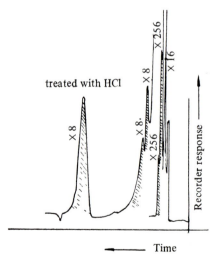

Fig. 93   Head space analysis of a polyacetal treated with HCl.

purpose.[15-18] Heins *et al.* have used this method to identify compounds in the vapour phase above tea and coffee with the aid of a 50 m polypropylene glycol capillary column.[19] The volatile constituents of the flavour of rum have been identified by Liebich *et al.*[20] The most important requirement for quantitative head space analysis is reliable performance of the equipment. The operating conditions must be more reproducible and stable than for any other comparable method, and *e.g.* careful preparation and thermostatting of the sample in its container is essential for the equilibrium vapour pressure to be established. Care must be taken in sample collection and transfer so as to avoid changes in its composition due to condensation. It is also important to thoroughly clean the sampling unit in order to prevent residues being carried over with the subsequent sample.

Gottauf has described a variation of quantitative head space analysis for the determination of traces of volatile organic compounds in water.[21] This involved concentration of the compounds present in the head gases in a cooled tube packed with an adsorbent, before actual gas chromatographic analysis. By taking special precautions, loss of material by adsorption or absorption was prevented, and the limit of detection was below $10^{-8}\%$ with a relative standard deviation of about $\pm 0.05\%$.

Even manual sampling with syringes from containers fitted with rubber caps involves possible errors which can significantly affect the reproducibility and accuracy of analysis.[8] For example, the type of rubber in the seal is very important. In the analysis of solids and substances producing aromas, Maier found that rubber stoppers absorb considerable amounts of the trace components being analysed. According to this author, this absorption cannot be avoided by heating the stopper or covering it with aluminium foil or Teflon film, while experiments with silvered screw caps also gave unsatisfactory results.[22] When analysing solids, it is therefore advisable to carry out a blank determination and allow for this in the results. On the other hand, when liquids are being analysed, no decrease in concentration of the trace components in the gas phase is observed, probably because they volatilize sufficiently rapidly from the liquid phase. Figure 94 shows the increase in amounts of various vapours absorbed in rubber stoppers as a function of time.

Davis has stated that this absorption in the material used to seal the sampling vessel head space is the main source of error, and depends on the molecular weight and molecular structure of the compound being analysed.[23] Thus, when a rubber stopper is used to seal the head space, within 30 minutes the concentration can decrease by 2% for ethylene, 7.6% for hexane and 21.9% for heptane. The loss is even greater for aldehydes, being 4.6% for propionaldehyde, 26.3% for valeraldehyde and 64.5% for heptanal. Davis therefore recommended that glass stopcocks be used since they eliminate this effect. Binder has drawn particular attention to the sources of error in sample preparation.[24] The large differences in rates of diffusion of various substances in the head space can lead to errors, especially when the equilibration period is short and the head space is large. This problem can be avoided by using an impeller for efficient mixing in the head space.

The advantages and disadvantages of sampling and injecting with a syringe are described in a large number of papers. Binder has presented an excellent survey and drawn attention to an automatic head space analyser which was proposed by Jentzsch et al.[9] and developed in collaboration with Machata for the analysis of blood alcohol, and which has subsequently found

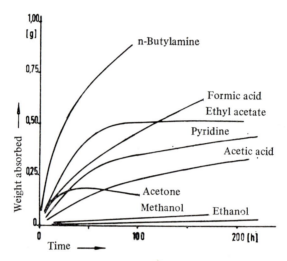

Fig. 94   Absorption of organic vapours in rubber stoppers at 23°C (from H. G. Maier, *J. Chromatog.*, **50**, 330 (1970)).

universal application in head space analysis.[25,26] This equipment (GC-Automat Multifract F40, Bodenseewerk Perkin-Elmer, Überlingen, Germany) functions completely automatically and avoids most of the drawbacks in the manual system mentioned above. The method of operation is as follows:

The sample is introduced into a serum bottle having a volume of about 30 ml. It is then hermetically sealed by means of a rubber septum and an aluminium foil, which is compressed onto the serum bottle with a special pair of pliers. A thermostat capable of holding 30 of these samples controls them at the same temperature, and they are then brought in succession to the automatic injector by means of a turntable.

At the beginning of an analysis, the turntable is raised and a hypodermic needle penetrates the rubber septum on the serum bottle. At the same time the controller regulated carrier gas pressure increases until it equals that at the column inlet. A fixed time is required for this after which the carrier gas valve closes. The excess pressure in the serum bottle is slowly released via the column, the volatile components in the sample being simultaneously carried into the column. At the end of this sampling operation the valve on the automatic injector closes and the carrier gas supply to the column is resumed. It is arranged that when the analysis is completed the turntable lowers and additional carrier gas is blown through the automatic injector via another valve. This removes the less volatile components which would otherwise be carried over into the next analysis. The duration of this flushing can be controlled as required.

When this sequence of operations is finished, the next sampling is started, and the instrument automatically stops after the 30 samples. Figure 95 is a schematic representation of the sampling system.

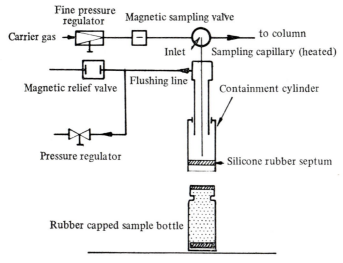

Fig. 95   Schematic diagram of the sampling system (from D. Jentzsch *et al.*, *Z. Anal. Chem.*, **236**, 98 (1968)).

The results of blood alcohol analyses carried out by this method are in good agreement with those obtained by the conventional Widmark method.[9]

Apart from the uncertainty caused by the use of rubber or plastic sealing caps, this automated method completely satisfies the instrumental requirements for quantitative head space analysis. However, there are still problems owing to the diversity of samples.

Gas chromatographic head space analysis enables the quantitative composition of the gas phase to be determined, but the result is initially only relative, since the main constituent, nitrogen (*i.e.* carrier gas), is not measured. Moreover, calculations of the concentration of a component 'i' in the sample from its amount in the gas phase are not necessarily valid, since a linear relationship between partial pressure and concentration can only be expected when solutions and gaseous mixtures are ideal.

If, as an example, a mixture of several volatile components is considered, then the total vapour pressure above the mixture is composed of the partial pressures of its components:

$$P_s = P_1 + P_2 \cdots + P_i = X_1 P_1{}^0 + X_2 P_2{}^0 \cdots + X_i P_i{}^0 \qquad (14)$$

where X_1, X_2 and X_i are the mole fractions and $P_1{}^0$, $P_2{}^0$ and $P_i{}^0$ are the vapour pressures of the pure components at the corresponding temperature.

It has been assumed that the partial pressures are sufficiently low to be identified with Dalton's partial pressures, *i.e.*

$$P_i = \frac{n_i RT}{v} \qquad (15)$$

so that the numbers of moles n_1, $n_2 \ldots n_i$ can be determined by simple chemical analysis of the vapour volume v.

However, there are two points to be considered:

1. Complications arise as soon as the pressure becomes so high that deviations from the gas laws occur.
2. Experience has shown that the dependence of the total pressure P_s and partial pressures P_1, P_2, ... P_i on the composition of the mixture varies somewhat from case to case, so that the following relations must be used:

$$P_s = \gamma_1 X_1 P_1{}^0 + \gamma_2 X_2 P_2{}^0 + \cdots \gamma_i X_i P_i{}^0 \qquad (16)$$

where γ_1, γ_2 and γ_i are the activity coefficients.

Four different situations can appear to varying degrees in head space analysis owing to diversity of the samples:

1. Linear dependence according to equation 14 when the substances do not mutually interact, *e.g.* hexane/heptane, O_2/N_2 or ethylene/propylene chloride mixtures. In these cases $\gamma = 1$.
2. According to equation 16, the total and partial pressures of the components in real mixtures exhibit deviations above and below a linear dependence on the mole fraction. If the attractive forces between the different kinds of molecules 1, 2, ... i are greater than those which would exist between molecules of the same kind 1, 2, or ... i if they were present as pure materials, then the partial vapour pressures will be lower than in an ideal mixture. This will give a negative deviation from Raoult's law. These attractive forces will be of electrostatic origin if the molecules possess permanent or induced dipoles which can lead to the formation of hydrogen bonds or other labile chemical bonds between the molecules. Thus, in the case of acetone + chloroform, a hydrogen bond is formed between the keto group of the acetone, which is a proton acceptor, and the chloroform, which is a proton donor. As a consequence, the partial vapour pressures of both liquids are lowered.

A further example of this type is aqueous hydrogen chloride. A sharp decrease in vapour pressure occurs as a result of intense interaction between H_2O and HCl with the formation of $(H_3O^+ + Cl^-)$. Up to a certain concentration, it is absolutely impossible to detect HCl pressure. Only pure water vapour appears over dilute solutions, and only HCl vapour over very concentrated solutions.

3. On the other hand, if the attractive forces between molecules of the pure components 1, 2, ... i are greater than those between the different kinds of molecules in the mixture they attempt to displace each other from the mixture, so that an increase in their partial vapour pressures occurs (positive deviation from Raoult's law).

As a rule, this behaviour occurs in mixtures of liquids containing molecules having polar groups and molecules having non-polar groups $e.g.$ ethanol/heptane.

4. Finally, there are also mixed types of vapour pressure curves in which convex and concave curvature occurs, $e.g.$ H_2O/pyridine.

Clearly in cases 2 to 4 the partial pressure may not vary linearly with, or may be independent of, the concentration, in the higher concentration regions, and thus head space analysis then gives completely incorrect results or gives the same result for different concentrations.

The main significance regarding head space analysis is that more calibration work is needed than in ordinary gas chromatographic analysis. This can be complicated in the case of a polymer dispersion involving two physical states, $i.e.$ a mixture of several liquid components and a solid phase. However, these difficulties are surmountable to some extent, if the work is carried out at sufficiently high dilution, $i.e.$ by approaching the region of ideal mixtures. Figure 96 shows that, for low concentrations of acrylonitrile in water, the peak area is a linear function of concentration.[12]

Özeris and Basette have carried out quantitative studies at even lower concentration levels and established a linear relationship between peak height and concentration of organic components in the range 0·01–10 ppm for the analysis of head gases over aqueous solutions of alcohols, ketones, aldehydes, esters and sulphides as shown in Fig. 97.[27]

The different slopes of the lines in Fig. 97 depend on the type of compound, and define the limits of detection. While most of these compounds can be analysed down to 0·01 ppm, alcohols cannot be determined below 0·1 ppm, and methanol not even below 1 ppm. This is due to the low methanol concentration in the head space and probably also to the relatively low sensitivity of the FID to the compound.

It follows from these results that whenever possible very dilute solutions should be used. At the same time, the potential of head space analysis with regard to enrichment of components in the head gas should be exploited. There are several possible ways of doing this.

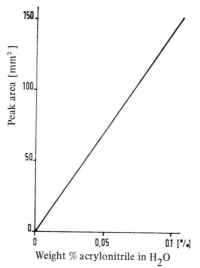

Fig. 96   Calibration of low concentrations of acrylonitrile in water for head space analysis.

One method of increasing the sensitivity of head space analysis over aqueous solutions consists of adding inorganic salts, *e.g.* anhydrous sodium sulphate. This has been reported to give a limit of detection of 0·01 ppm for carbonyl compounds.[13] Kepner *et al.* have obtained up to a sevenfold increase by adding ammonium sulphate or sodium chloride until the dilute aqueous sample was saturated. The equilibrium vapour pressure was established by swirling the solution at constant temperature.[28] A linear dependence of peak height on the amounts of various alcohols and esters was also obtained by these authors, who used an internal standard. Jentzsch *et al.* showed that a significant increase in sensitivity towards *t*-butanol in aqueous solution

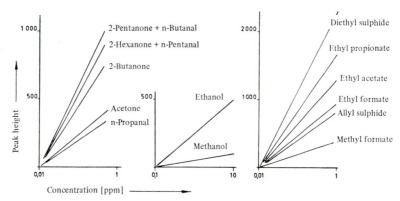

Fig. 97   Peak heights as a function of concentration of various organic substances (from S. Özeris and R. Basette, *Anal. Chem.,* **35,** 1091 (1963)).

can be obtained by adding calcium carbonate.[29] The limit of detection was reported to be 0·05 ppm.

A change in the equilibrium vapour pressure on decreasing the solubility in aqueous solutions naturally also occurs with the other components present in the sample, and this must be taken into account in their analysis. Their concentration varies greatly from sample to sample and can give rise to considerable difficulties on quantitative analysis. Thus, for example, high concentrations of ethanol affect the solubility, and consequently the vapour pressure, of other compounds present in lower concentration. The same effect is also produced by *e.g.* the high concentrations of carbon dioxide present in the head space analysis of beer and lemonade. It has been shown experimentally that peak heights for isoamyl acetate, isobutanol and isoamyl alcohol in aqueous solutions are smaller than in beer. On the other hand, some compounds are less influenced by the effects mentioned above. Thus, ethyl caproate has the same peak height in water as in rum. This illustrates how difficult and time-consuming quantitative head space analysis can be, since it necessitates a special calibration for each product being analysed.[28]

According to Jentzsch *et al.*, for non-aqueous solutions enrichment of one component relative to another presupposes differences in the shape of vapour pressure curves.[29] At the same time, the sample vessel temperature must be so chosen that the difference in vapour pressures is as high as possible. Thus, for example, with toluene containing 1% benzene, conventional sampling of the liquid gives chromatogram *A*, and sampling from the head space gives chromatogram *B* (Fig. 98).

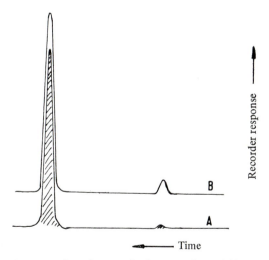

Fig. 98   Chromatograms of a mixture of toluene and 1 vol % benzene. Chromatogram *A*: conventional sampling, Chromatogram *B*: sampling from the head space at 40°C (from D. Jentzsch *et al., Angew. Gaschromatographie,* (1967) part 9, Bodenseewerk Perkin-Elmer).

Whereas, the ratio of the two components (benzene to toluene) in the liquid phase is 1:100, based on the composition of the sample, it follows from chromatogram *B* that head space analysis gives a ratio of 4·3:100, *i.e.* the benzene content of the sample arriving at the gas chromatograph at 40°C has increased by a factor of 4·3. In this way, the limit of detection of liquid components (in this case benzene) is reduced, *i.e.* even smaller amounts can be detected. Very low concentration levels are involved in blood alcohol analysis but, according to Machata, addition of an internal standard makes it possible to perform very accurate analyses.[26] Hauck and Terfloth have investigated the sources of error in this kind of blood alcohol analysis when carried out automatically.[30] They found that increasing the temperature of the sample container by 1 deg C produced equal increases in the peak heights of the ethanol and internal standard (butanol). The use of an internal standard can therefore be fully recommended in head space analysis. It also appears, in accord with observations on aqueous dispersions of polymers, that the ethanol peak is greater in blood than in aqueous ethanol samples having the same concentration. This parallelism results from the fact that in both cases two phases are present, one liquid and one solid. Hachenberg has carried out quantitative analyses of low concentrations of vinyl acetate, polymerized in methanol solution, demonstrating both the accuracy and reproducibility of analysis of dispersions by the head space method.[12] It was recommended that more concentrated solutions of vinyl acetate should suitably be diluted with methanol before analysis. In Table 3 the analytical

TABLE 3

Sample	Weight % vinyl acetate determined by titration	Weight % vinyl acetate determined by head space analysis				
		1	2	3	4	5
A	7·0	7·6	7·7	7·5	7·6	7·6
B	0·5	0·7	0·7	0·7	—	0·7
C	19·0	19·1	18·7	18·0	—	18·6

results for the head space method are compared with values obtained by the bromide–bromate titration for vinyl acetate in polyvinyl acetate, polymerized in methanol solution.

The quantitative head space analysis of solids is even more difficult as far as calibration is concerned. If the solid can be dissolved in a solvent then there is the possibility of analysing the head gas over the solution, but there are many cases in which this procedure is not applicable, *e.g.* tobacco, fruits, many polymers, etc. The technique employed in an example previously mentioned (*see* Fig. 89), where the styrene content of polystyrene was determined very accurately by gas chromatography, on dissolving the sample in methylene chloride and precipitating with methanol, may be used to calibrate head space analyses if an internal standard is added. Figure 99 shows a calibration from this kind of measurement which can form the basis of the quantitative head space analysis of styrene in polystyrene.[12]

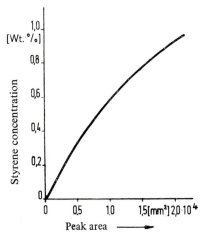

Fig. 99   Relationship between weight % styrene in polystyrene and peak area (styrene) for the head space analysis.

Rohrschneider has shown experimentally that equilibration between polystyrene granules and the gas phase is established too slowly for a gas phase analysis.[31] He therefore recommended that this analysis should be carried out with a solution of the polymer in dimethylformamide, since equilibrium

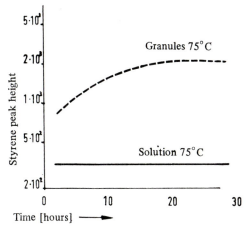

Fig. 100   Equilibration of styrene monomer between the gas phase and polystyrene granules or polystyrene solution, in the head space analysis of polystyrene (from L. Rohrschneider, *Z. Anal. Chem.*, **255**, 345 (1971)).

is then established sufficiently rapidly, *i.e.* in a maximum of two hours. In Fig. 100, the approach to equilibrium for polystyrene granules is compared with the behaviour of polystyrene solution. The gas chromatograms of the gas phase above polystyrene solution and a corresponding standard sample are presented in Fig. 101.

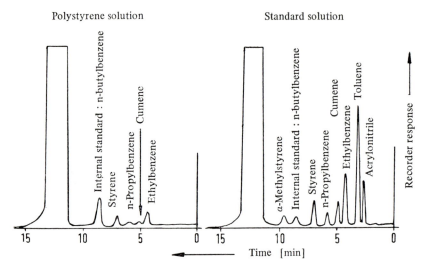

Fig. 101   Gas chromatogram of the gas phase above a polystyrene solution and above a standard sample (from L. Rohrschneider, *Z. Anal. Chem.*, **255**, 345 (1971)).

According to Rohrschneider, the limit of detection of styrene for this analysis, which may also be carried out with the automatic head space analyser F40 (Perkin-Elmer Ltd., Bodenseewerk) is 10 mg/kg of polystyrene. The time required for an analysis can be reduced to 6 minutes by calculating the results on a computer.

References
1. MACKAY, D. A.M., LANG, D. A. and BERDICK, M., *Anal. Chem.* **33**, 1369 (1961).
2. BROWN, D. S., BUCHANAN, J. R. and HICKS, J. R., *Proc. Amer. Soc. Hort. Sci.,* **88**, 98 (1966).
3. ROMANI, R. J. and KU, L. L., *J. Food Sci.,* **31**, 558 (1966).
4. DRAWERT, F., HEIMANN, W., EMBERGER, R. and TRESSL, R., *Chromatographia,* **2**, 57 (1969).
5. WEURMAN, C., *Food Technol.,* **15**, 531 (1961).
6. ELKINS, E. R., KIM, E. S. and FARROW, R. P., *Food Technol.,* **23**, 1419 (1969).
7. RASCEKH, J. A., *Dissertation Abstr.,* **30** No. 3, 1189B, September 1969, Univ. Maryland.
8. MIETHKE, H. *Deut. Lebensm.-Rundschau,* **65**, 379 (1969).

9. JENTZSCH, D., KRÜGER, H., LEBRECHT, G., DENCKS, G. and GUT, J., *Z. Anal. Chem.*, **236**, 112 (1968).
10. KOLB, B., WIEDEKING, E. and KEMPKEN, B., *Angewandte Gas-chromatographie*, Bodenseewerk Perkin-Elmer, Part 11–11E (1968) p. 5.
11. KIERMEIER, F. and STROH, A., *Z. Lebensm.-Untersuch. -Forsch.*, **141**, 216 (1969).
12. HACHENBERG, H., Bodenseewerk Perkin Elmer, *TIPS* 41 *GC*, April 1970.
13. BASETTE, R. ÖZERIS, S. and WHITNAH, C. H., *Anal. Chem.*, **34**, 1540 (1962).
14. PALO, V., *Chromatographia, 4*, 55 (1971).
15. HENNEBERG, P. and SCHOMBURG, G., *Z. Anal. Chem.*, **211**, 55 (1965).
16. TERANISHI, R., CORSE, J. W., McFADDEN, W. H., BLACK, D. R. and MORGAN, A. I., *J. Food Sci.*, **28**, 478 (1963).
17. DAY, E. A. and LIBBEY, L. M., *J. Food Sci.*, **29**, 583 (1964).
18. McFADDEN, W. H., TERANISHI, R., CORSE, J. W., BLACK, D. R. and MON, T. R., *J. Chromatog.*, **18**, 10 (1965).
19. HEINS, J. TH., MAARSE, H., TEN NOEVER DE BRAUW, M. C. and WEURMAN, C., *J. Gas Chromatog.*, **4**, 395 (1966).
20. LIEBICH, H. M., KÖNIG, W. A. and BAYER, E., *J. Chromatog. Sci.*, **8**, 527 (1970).
21. GOTTAUF, M., *Z. Anal. Chem.*, **218**, 175 (1966).
22. MAIER, H. G., *J. Chromatog.*, **50**, 329 (1970).
23. DAVIS, P. L., *J. Chromatog. Sci.*, **8**, 423 (1970).
24. BINDER, H., *Z. Anal. Chem.*, **244**, 353 (1969).
25. MACHATA, G., *Blutalkohol, 4*, 3 (1967).
26. MACHATA, G., *Mikrochim. Acta*, 262 (1964).
27. ÖZERIS, S. and BASETTE, R., *Anal. Chem.*, **35**, 1091 (1963).
28. KEPNER, R. E., MAARSE, H. and STRATING, J., *Anal. Chem.*, **36**, 77 (1964).
29. JENTZSCH, D., KRÜGER, H. and LEBRECHT G., *Angewandte Gas-chromatographie, 19*, (1967) Part 9, Bodenseewerk Perkin Elmer.
30. HAUCK, G. and TERFLOTH, H. P., *Chromatographia, 2*, 309 (1969).
31. ROHRSCHNEIDER, L., *Z. Anal. Chem.*, **255**, 345 (1971).

2.16. Gases in Liquids

Special precautions are necessary in the collection of samples for determination of traces of gases dissolved in liquids (*see* 1.331). Moreover, sample injection can only be carried out with a syringe if the differences in boiling points between the gaseous components being analysed and the liquid phase are relatively small, and if the gases are readily soluble in the liquid. If inert gases have to be determined in liquid systems, *e.g.* H_2, O_2, CH_4, CO and CO_2 in water, N_2 in blood, or O_2 and N_2 in blood or hydrocarbon fractions, then this method of sample collection cannot be recommended since fractionation can occur as the sample is drawn into the syringe.[1–5] This degassing process, which is caused by the reduced pressure produced when the syringe plunger is raised, can occur to such an extent that quantitative analysis is impossible. Also, when determining traces of inert gases, very large samples must be used since only the thermal conductivity detector can be employed. For this reason, the gaseous constituents must be separated from the liquid, and various methods have been developed for this purpose. Thus, the carrier gas can be passed over the sample containing the gas while it is being agitated.[6] However, the disadvantage is that the gases are not removed instantaneously, and so an indication of the true concentration of the gaseous components in the sample can only be obtained from repeated analyses. Another method employs a by-pass to equilibrate the sample with the carrier gas. This technique has the disadvantage that, when the by-pass is switched into the normal carrier gas stream, a violent base line fluctuation is produced on the chromatogram.[7] Another method consists of using the van Slyke apparatus.[8,9] Natelson and Stellate have developed an apparatus in which the solvent is constantly stirred under vacuum, which enables the gaseous constituents to be completely separated from the solvent.[5] Ford determined oxygen dissolved in hydrocarbons with the aid of a Teflon/ stainless steel sampling valve, which permitted sample extraction over a wide range of temperatures and pressures differing from those of the carrier gas system.[10] An electron capture detector was employed. Sampling of liquids containing gases can also be carried out by sealing the samples in glass ampoules which are subsequently crushed in the gas chromatograph. This is the best, and the fastest, method of separating gaseous constituents from the liquid. Pichuzhkin used this method to determine hydrogen and C_1–C_4 hydrocarbons in heptane.[11] Analysis was carried out with two gas chromatographs in series. The gaseous components were separated from the heptane on a 2 m dinonyl phthalate column. The C_2 and C_4 hydrocarbons were subsequently analysed on alumina coated with NaOH, and a molecular sieve column was used for the methane and hydrogen. A thermistor detector was employed, and the limit of detection was reported to be about 2 vol ppm for C_2 hydrocarbons.

Separation of gaseous constituents from a solution can also be conducted in a separate apparatus. This has the advantage that high enrichment factors can be achieved. Thus, a conventional, or preferably a preparative Janák apparatus (*see* 1.341), can be used to separate and concentrate inert gases and C_1–C_4 hydrocarbons from higher boiling hydrocarbons.[12] This method cannot, of course, be used with gases which react with KOH. Furthermore, the enrichment factor is limited to a certain extent due to the capacity of the column for the liquid being analysed. In many cases, this may be avoided by 'stripping' the solution with CO_2 instead of separating the gaseous components from the liquid in the chromatographic column. An apparatus of this type, which can be used for the determination of traces of oxygen in methanol, is shown in Fig. 102.[13]

The function of this apparatus is to flush an accurately measured amount of methanol with completely air-free CO_2. The inert gas fraction containing the oxygen is collected in the nitrometer whilst methanol carried over with it is absorbed by aqueous KOH. The amount of oxygen is determined in the subsequent gas chromatographic analysis of the inert gas fraction and,

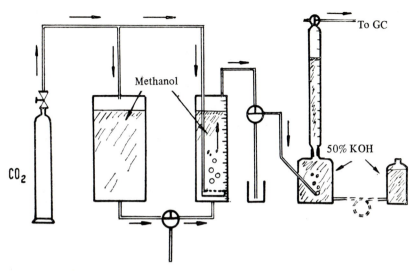

Fig. 102   Apparatus for determination of gases in liquids.

by relating it to the quantity of methanol, the concentration of oxygen may be calculated as follows:

$$\text{ppm by weight} = \frac{A(\text{mg})10^6}{B(\text{mg})}$$

where A is the weight of oxygen in mg determined by the g.c. analysis, and B is the weight of methanol used in mg.

The data in Table 4 demonstrate the reproducibility of the 'stripping' method.

TABLE 4

Volume of methanol (ml)	Volume of inert gas (ml)
350	26
350	24
350	27
350	23
350	28
350	25
350	26
350	25
350	24
350	27
350	23

This method, which can of course be applied to other problems by using a different liquid in the nitrometer, has a limit of detection of less than 1 ppm.

References

1. WALKER, J. A. and FRANCE, E. D., *Analyst,* **94,** 364 (1969).
2. ROPARS, J., *Chim. Anal.* (Paris), **50,** 641 (1968).
3. WILDE, B. E. and HODGE, F. G., *Ind. Eng. Chem., Prod. Res. Develop.,* **8,** 408 (1969).
4. IKELS, K. G., *J. Gas Chromatog.,* **3,** 359 (1965).
5. NATELSON, S. and STELLATE, R. L., *Anal. Chem.,* **35,** 847 (1963).
6. TAYLOR, B. W. and PRESSAU, J., *J. Physiologist,* **2,** 114 (1959).
7. HAMILTON L. H., *Ann. N.Y. Acad. Sci.,* **102,** 15 (1962).
8. LUKAS, D. S. and AYERS, S. M., *J. Appl. Physiol.,* **16,** 371 (1961).
9. RAMSEY, L. H., *Science,* **129,** 900 (1959).
10. FORD, P. T., *Anal. Chem.,* **41,** 393 (1969).
11. PICHUZHKIN, V. J., *Khimiya,* 25 (1966).
12. KUDRYAVTSEVA, N. A. and SHCHIPANOVA, A. I., *Zavod. Lab.,* **34,** 146 (1968).
13. HACHENBERG, H., unpublished work.

2.2. LIQUID HYDROCARBONS

Choice of a detector for trace analysis of hydrocarbons in mixtures of other hydrocarbons, or determination of the purity of individual hydrocarbons, does not present any problems since the sensitivity of the FID is entirely adequate. In both cases, the selectivity of the chromatographic column is of much greater importance. The problem of resolution becomes very difficult when it is required to analyse traces of materials having a wide range of boiling points in hydrocarbon mixtures. In this connection, the earlier discussion may be recalled of the analysis of the higher acetylenes in C_1–C_5 hydrocarbons (*see* 2.13). Mixtures of liquid hydrocarbons usually consist of petroleum fractions containing 150–250 individual constituents, and the analytical problem is frequently determination of traces of aromatics and unsaturated hydrocarbons in these mixtures. Liquid chromatography (fluorescent indicator technique) may be used at the percentage level, but not at the ppm level since it is not sufficiently sensitive. The same is true of infrared spectroscopy. There are no specific detectors available for unsaturated hydrocarbons and aromatics if the use of ultraviolet spectroscopy for determination of the total aromatic content is disregarded. It is therefore necessary to carry out very selective separations.

DETERMINATION OF TRACES OF AROMATICS

Aromatics may be separated from paraffins of the same boiling point by using selective columns. For this separation, Durrett recommends Carbowax 400 as stationary phase since the retention times of the aromatics are increased to such an extent compared with those of the paraffins that benzene, which boils at 80°C, is eluted together with the C_{10} paraffins (boiling points about 170°C), as can be seen from Fig. 103.[1]

The gas chromatographic determination of aromatics in aviation fuel with this column is shown in Fig. 104.

Another selective phase for this separation is 1,2,3,4,5,6-hexakis(2-cyanoethoxy)hexane.[2] With this stationary phase, it is possible to elute benzene after the C_{11} n-paraffin. Figure 105 shows a typical analysis of 10 ppm by weight each of benzene, toluene, ethylbenzene, *m*-xylene, *p*-xylene, cumene, *o*-xylene, *m*-di-isopropylbenzene and *p*-di-isopropylbenzene in a

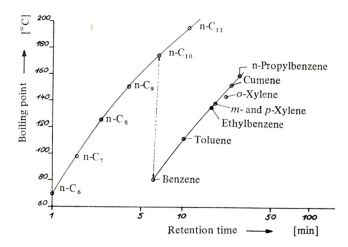

Fig. 103 Selectivity of Carbowax 400 for determination of aromatics in paraffins (from L. R. Durrett, *Anal. Chem.*, **32,** 1393 (1960)).

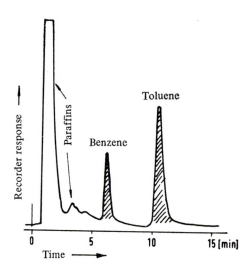

Fig. 104 Determination of aromatics in aviation fuel with Carbowax 400 as stationary phase (from L. R. Durrett, *Anal. Chem.*, **32,** 1393 (1960)).

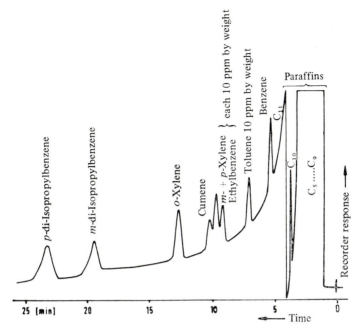

Fig. 105 Selective determination of aromatics in paraffins with 1,2,3,4,5,6-hexakis(2-cyanoethoxy)hexane as stationary phase. (Column length: 5 m, internal diameter: 4 mm, column temperature: 75°C, FID, Sample: 0·001 ml.)

mixture of C_1 to C_{11} paraffins. It is possible to detect 1 ppm by weight of each of the components.

Small amounts of higher boiling components often interfere with the analysis because they have the same retention time as the aromatic compound being determined, and cannot be distinguished from it on the chromatogram. Nevertheless, they may be differentiated by sulphonating the sample (ASTM-D-1019-62) and then repeating the gas chromatographic analysis. This can give rise to three situations:

(a) the peaks which had been identified from retention times as due to aromatic compounds may disappear completely after sulphonation. In this case they are entirely from aromatic compounds.

(b) if the peaks have equal areas before and after sulphonation then they do not result from aromatic compounds, and are therefore from paraffins or naphthenes.

(c) the areas of the peaks after sulphonation may be smaller than on the original trace. This will occur when paraffins, naphthenes and aromatics having the same retention time are present. The concentration of aromatic compounds has to be deduced from the difference in areas of the peaks before and after sulphonation.

In the determination of aromatic compounds, Lebbe *et al.* carried out a pre-separation on a non-polar column (1·5 m Apiezon L).[3] Selective separation was subsequently achieved on tetrakis(2-cyanoethyl)pentaerythritol as the stationary phase. Lijinsky *et al.* have determined traces of polynuclear hydrocarbons in petroleum and coal tar on a column of glass beads coated with SE-30 and operated at temperatures between 160°C and 210°C.[4] Before analysis, the aromatic compounds were concentrated by distribution between two suitable solvents, *e.g.* cyclohexane–nitromethane. A limit of detection of 0·1 ppm was obtained with a strontium-90 ionization detector.

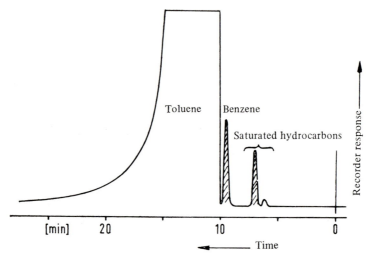

Fig. 106   Determination of traces of hydrocarbons in toluene (from F. A. Fabrizio *et al., Anal. Chem.,* **31**, 2060 (1959)).

The converse analytical problem, determination of saturated hydrocarbons in aromatics, also requires specially selective columns. However, as far as the sensitivity of detection is concerned, the task is simpler here since the traces of paraffins being analysed can be eluted before the main components. For example, when 2,3,7-trinitro-9-fluorenone is used as stationary phase, saturated hydrocarbons up to C_{10} may be determined in toluene, and also in benzene, with a combined picric acid, di-n-butyl phthalate and di-n-decyl phthalate stationary phase as shown in Figs. 106 and 107.[5]

For the gas chromatographic analysis of crude benzene, Jaworski and Bobinski used a combined column (diameter 6 mm) consisting of a 2 m long section containing polypropylene glycol adipate and a 2 m section of polyethylene glycol phthalate. The column temperature was 130°C and the stationary phases were supported on porous brick (activated for 5 hours at 350°C). The individual components of crude benzene which were identified emerged from the column in the following order: n-pentane, n-hexane,

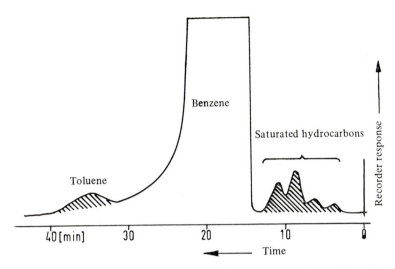

Fig. 107   Determination of traces of hydrocarbons in benzene (from F. A. Fabrizio
et al., Anal. Chem., **31,** 2060 (1959)).

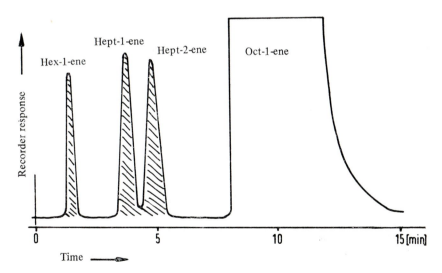

Fig. 108   Determination of unsaturated hydrocarbons in saturated hydrocarbons
(from C. G. Scott and C. S. G. Phillips, *Nature,* **199,** 66 (1963)).

n-heptane, carbon disulphide, n-octane, n-nonane, benzene, thiophen, toluene, ethylbenzene, *o*-xylene, *m*- + *p*-xylene, styrene, dicyclopentadiene, 1,2,4- and 1,2,3-trimethylbenzene, indane and indene.[6]

According to Scott and Phillips, traces of alkenes may be determined in alkanes by means of a displacement technique.[7] The sample being analysed is passed through a column of alumina modified with silver nitrate, from which the alkanes are then eluted with nitrogen. The hexenes and heptenes are subsequently eluted with nitrogen saturated with octene-1 vapour (*see* Fig. 108).

References
1. DURRETT, L. R., *Anal. Chem.*, **32,** 1393 (1960).
2. HACHENBERG, H., unpublished work.
3. LEBBE, J., DUCROS, M. and GUENIER, J. P., *Occupational Health Rev.,* **19,** 3 (1967).
4. LIJINSKY, W., DOMSKY, I., MASON, G., RAMAHI, H. Y. and SAFAVI, T., *Anal. Chem.*, **35,** 952 (1963).
5. FABRIZIO, F. A., KING, R. W., CERATO, C. C. and LOVELAND, J. W., *Anal. Chem.*, **31,** 2060 (1959).
6. JAWORSKI, M. and BOBINSKI, J., *Chem. Analit.* (Warsaw), **9,** 1003 (1964).
7. SCOTT, C. G. and PHILLIPS, C. S. G., *Nature,* **199,** 66 (1963).

2.3. DETERMINATION OF TRACES OF WATER

The determination of traces of water in gases, liquids and solids is one of the most important, and at the same time most difficult, problems of trace analysis, since water is universally present, and because it is highly polar it adheres to all surfaces.

The best known means of determination, which can still be used in the majority of cases, is the Karl Fischer method.[1] However, this technique is known to have some disadvantages, and *e.g.* compounds which react with iodine cannot be analysed. Moreover, analysis of gases and solids is very time-consuming since it must be preceded by concentration or extraction stages. Also, the lower detectable limit is generally in the region of 1–5 ppm.

An alternative method for continuous determination of small amounts of water in gases employs an electrochemical instrument, in which water reacts with phosphorus pentoxide which is regenerated electrolytically, the current required being taken as a measure of the water concentration. Apart from the fact that it can only be used for gases, this method also has the disadvantage that it cannot be employed with certain polymerizable compounds. Thus, it cannot be used to determine water in propylene, since the phosphorus pentoxide-coated electrode becomes covered with a layer of polymer which renders the cell unserviceable. This problem may be overcome by concentrating the water present in the gas sample on a polyethylene glycol column, and then using a carrier gas to elute it into the electrolytic cell.[2] Calibration for small amounts of water vapour can be carried out by adsorption and absorption techniques at low temperatures but these are not recommended for gas analyses in view of the poor reproducibility.

In order to overcome the above problems there have been many attempts to determine traces of water by gas chromatography. This is not easy since water is a polar compound and not readily eluted from normal columns. Apart from memory effects, the water peak is usually asymmetric and shows considerable tailing, which makes it very difficult to determine trace amounts in the region below 100 ppm. The flame ionization detector, generally employed for trace analysis, cannot be used in this case since it does not respond

133

to water. Only the relatively insensitive thermal conductivity detector can be used for this compound.

The two basic possibilities of avoiding the above problems are summarized in a paper by Neumann.[3] *Indirect determination* of traces of water involves conversion into other compounds to which the FID is sensitive, and *direct determination* is generally carried out with a thermal conductivity detector after pre-concentration, or directly with certain ionization detectors.

Indirect determination: In the indirect method, water is reacted with certain compounds, and the reaction products are determined gas chromatographically. For example, water forms acetylene with calcium carbide which can be measured with high sensitivity by gas chromatography.[4-7]

$$CaC_2 + 2H_2O \longrightarrow Ca(OH)_2 + C_2H_2$$

This obvious and simple reaction for the detection of water often involves complications which have to be taken into account. For example, in the reaction with calcium carbide localized heating can occur and cause decomposition or polymerization of the acetylene formed, while any calcium oxide produced can give rise to further problems. In addition, side reactions can occur with the other components present.

Goldup and Westaway have evaluated the conditions required for using this method to determine water in propane and butane.[8] The optimum temperature for the reaction was reported to be 40°C. The calcium carbide powder was conditioned by passing dry nitrogen over it at 140°C, and the reactor was 6–8 cm long and had an internal diameter of 3 mm. For accurate determination of the water, the reactor temperature must be controlled to ± 0.1 deg C, and with careful attention to these conditions it should be possible to determine down to 0.01 vol ppm of water in gases.

Investigations of the possible side reactions of calcium carbide showed that there is no interference from methanol at a concentration of 3500 vol ppm and from other alcohols at concentrations of 600 vol ppm. Knight and Weiss confirmed these results but found that hydrogen sulphide and carbonyls do interfere.[9]

According to these authors, the Karl Fischer determination of water vapour in butadiene gives results in good agreement with those from the calcium carbide method:

Sample	Karl Fischer	CaC_2 Method
1	3 ppm	3–4 ppm
2	16 ppm	14 ppm
3	25 ppm	22 ppm

The error is about 10% at 20 ppm. Kung *et al.* have developed a special Pyrex glass reactor in which conversion of water into acetylene was carried out at 220°C, the calcium carbide being mixed with glass beads.[10] Kaiser has combined the conversion of water into acetylene with the use of reversion g.c. and attained a limit of detection of 10^{-2} ppb.[11] Since he worked with a continuous flow system, the errors mentioned above for the reaction with calcium carbide were minimized.

The formation of hydrogen in the metathetic reaction with calcium hydride offers another possible method of determining traces of water.[4,12]

$$CaH_2 + 2H_2O \longrightarrow Ca(OH)_2 + 2H_2$$

Since the amount of gas evolved in this reaction is twice that from the decomposition of calcium carbide, this method is preferable when thermal conductivity detectors are employed. Moreover, there are fewer possible side reactions than with calcium carbide. Gelezunas has reported that side reactions do not occur between calcium hydride and oxygen, butane, carbon dioxide and ethylene, but slow reactions were observed in the case of ethanol and acetone.[13] The use of calcium hydride labelled with tritium offers additional advantages regarding the sensitivity, since a radiological detector can be used. This can give a limit of detection of $10^{-7}\%$.

Another reagent which should be mentioned for the indirect determination of water is lithium aluminium hydride. Its reaction with water

$$LiAlH_4 + 4H_2O \longrightarrow LiOH + Al(OH)_3 + 4H_2$$

produces an even greater amount of gas to be measured.[14] Berezkin et al. have used sodium aluminium hydride and measured the corresponding yield of hydrogen with a thermal conductivity detector.[15] They were able to determine 2–3 ppm of water in a 10 ml sample of hydrocarbon.

The reaction between lead tetra-acetate and water, in which 4 moles of acetic acid are formed:

$$Pb(CH_3COO)_4 + 2H_2O \longrightarrow PbO_2 + 4CH_3COOH$$

can also be used for the gas chromatographic determination of traces of water.[16]

Other possible methods are based on the reactions of water with alkyl-magnesium halides, sodium or potassium amalgams, magnesium nitride, or with dimethoxypropane which forms acetone:

$$CH_3-\underset{\underset{OCH_3}{|}}{\overset{\overset{OCH_3}{|}}{C}}-CH_3 + H_2O \longrightarrow CH_3-\overset{\overset{O}{\|}}{C}-CH_3 + 2CH_3OH$$

in the presence of methanesulphonic acid as a catalyst.[17,18] This method was originally developed for the infrared spectroscopic determination of water (measurement of the carbonyl band at 5·87 μm) at the higher concentration levels, but it is particularly well suited to the gas chromatographic determination of traces of water in the form of acetone.[19] The decrease in the dimethoxypropane peak can also be measured, but of course this can only be used for higher water concentrations. It should be noted that basic compounds can interfere by reacting with the methanesulphonic acid and should therefore be removed prior to the analysis. Martin and Knevel have used tetrahydroxyethylethylenediamine as stationary phase to separate the acetone and dimethoxypropane.[20] Comparison with the Karl Fischer method showed that the DMP method can be more accurate and faster. A chromatogram showing the separation of DMP, acetone and methanol on a 2·4 m column of tetrahydroxyethylethylenediamine is reproduced in Fig. 109.

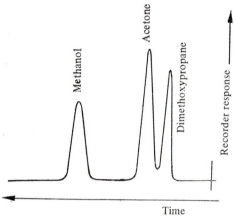

Fig. 109   Determination of water with dimethoxypropane (from J. H. Martin and A. M. Knevel, *J. Pharm. Sci.,* **54**, 1464 (1965)).

There are numerous applications of the indirect methods mentioned above, and the sensitivity with which the reaction products formed from water can be detected is of the highest order.

Direct determination: Quantitative analyses, particularly on a routine basis, are often insufficiently reproducible, however, because not enough attention is paid to side reactions.[4] For this reason, many attempts have been made to determine traces of water by direct methods. As it is not possible to employ an FID for this purpose, with a thermal conductivity detector relatively large samples must be used or pre-concentration must be undertaken.[21] Since tailing and memory phenomena occur in the analysis of water, there is only a relatively small choice of suitable column packings. The limit of detection is unsatisfactory for many of the direct methods of analysis, and, for example, it has been reported to be as high as 2000 ppm. However, Aubeau *et al.* have detected 100 ppm by using polyethylene glycol 1500 on Teflon powder as the solid support.[4] By concentrating the water on a precolumn containing the same packing a limit of 1 ppm can be obtained. An amount of 20 ppm of water can be detected on Teflon which has been coated with 15% tetrahydroxyethylene diamine.[23]

The determination of traces of water in organic solvents has been discussed in a paper by Beresnev *et al.*[24] The quantitative calibration of trace amounts of water was carried out in the region of 10 ppm in isoamyl alcohol. The column packing used was Apiezon L and polyethylene glycol 400 on Teflon. Separation of air and water is a problem in this analysis, and it also occurs in the determination of water vapour in air on a Carbowax 20 M column.[25]

With the column packings so far mentioned for direct determination of water, adsorption and long elution times are clearly disadvantages for trace analysis. Teflon powder is generally used as an inert support but it is not easy to handle owing to electrostatic charging. Alternatively, conventional solid supports may be used, but these must be deactivated before being coated with the stationary phase. Porous polymers based on styrene-divinyl-benzene (Porapak) are particularly suited to the determination of water since

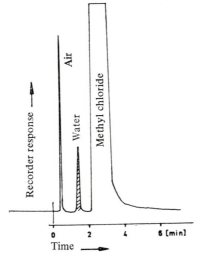

Fig. 110   Direct determination of traces of water in methyl chloride on a 1·8 m Porapak Q column at 98°C (from O. L. Hollis and W. V. Hayes, *J. Gas Chromatog.*, **4**, 236 (1966)).

they give a very sharp peak having a short elution time. They have been used to determine traces of water in hydrocarbons, alcohols, glycols, chlorinated hydrocarbons, ammonia and acids.[26] Figure 110 shows the determination of traces of water in methyl chloride on Porapak Q, and Fig. 111 illustrates the direct determination of water on Porapak T.

Neumann has reported excellent results for the direct determination of water in gases and liquids with Porapak R.[3] A limit of detection of 1 ppm of

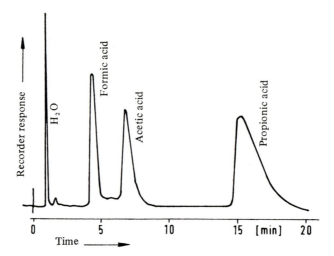

Fig. 111   Direct determination of traces of water in organic acids on a 1 m Porapak T column at 169°C (from O. L. Hollis and W. V. Hayes, *J. Gas Chromatog.*, **4**, 236 (1966)).

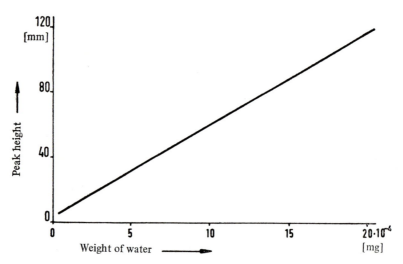

Fig. 112   Peak height as a function of concentration of water in the analysis of liquids (from G. M. Neumann, *Z. Anal. Chem.*, **244**, 302 (1969)).

water was obtained with a thermal conductivity detector. Calibrations showed that a linear dependence of peak height on concentration of water is established both for gases and liquids (*see* Fig. 112).

In the region of 100 ppm of water, the error is $\pm 8\%$ and at levels below 10 ppm it is $\pm 20\%$. This linear dependence was confirmed by Cotton *et al.* who used Porapak Q for the gas chromatographic determination of water in chlorophyll.[27] They found that solutions of water in benzene, containing acetone as internal standard, cannot be stored in ordinary volumetric flasks fitted with glass stoppers since the water concentration continuously increases owing to diffusion from the atmosphere. It was therefore recommended that such standard solutions should be kept in glass flasks having long 1 mm internal diameter necks and fitted with Teflon stoppers. If these precautions are observed then the gas chromatographic results are in good agreement with Karl Fischer titrations. Hogan *et al.* have recommended methanol as an internal standard in the determination of traces of water, and have also considered Porapak Q to be the most suitable column packing.[28] The standard

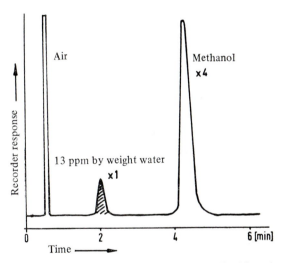

Fig. 113 Direct determination of traces of water in liquids using methanol as internal standard (from J. M. Hogan *et al.*, *Anal. Chem.*, **42**, 250 (1970)).

deviation was ± 0.6 ppm in the region of 10 ppm and ± 4 ppm at 50 ppm. A limit of detection of 0.1 ppm can be obtained with a thermal conductivity detector incorporating WX (Gow-Mac) detector elements. Figure 113 illustrates the direct determination of traces of water with methanol as internal standard.

In spite of thorough preconditioning of these crosslinked polymers (Porapak Q and R) by washing with methanol, water, aqueous hydrochloric acid (dilution 1:4) and acetone, it is found that memory effects can occur the first time a sample is injected. Kaiser recommends carbon molecular sieve for the gas chromatographic determination of traces of water.[11] Since carbon has a very low polarity compared with water, it is eluted very rapidly as a sharp symmetrical peak. According to the author, this material may be used for the direct determination of 20 ppm water in gases, and down to 0·2 ppm in liquids and fusible solid materials.

The limits of detection mentioned above were obtained with conventional thermal conductivity detectors. However, there are now more sensitive detectors available which enable even lower limits of detection to be attained in the direct determination of water. The helium and radiation ionization detectors generally appear to be promising.[29] Thus, if commercial neon containing 0·05 vol % argon, 0·1 vol % helium, 0·04 vol % oxygen, 0·02 vol % nitrogen and 0·001 vol % carbon dioxide is used as the carrier gas, it is possible to reach limits of detection which can certainly not be improved on by using highly purified carrier gases.[30]

When traces of water are being measured in gases, special attention should be paid to sampling, which, as for other highly polar compounds such as methanol or hydrogen sulphide, should on no account be a discontinuous process. Whenever possible the sample should be collected continuously in-line from the gas stream to be analysed. Even this can give rise to problems and, for example, the sampling valve and sample loop must be heated, and preferably should not be made of metal, except perhaps aluminium, but from a strongly water-repelling material such as PTFE.[31] In this case, Kaiser considers that the coupling lines from the sampling unit right through to the detector should be made of silica or pure aluminium, and that the temperature of the whole gas chromatographic unit should be maintained constant to $\pm 0·1 - \pm 0·01$ deg C.[11]

References
1. FISCHER, K., *Angew. Chem.,* **48,** 394 (1935).
2. British Patent 1,076,899.
3. NEUMANN, G. M., *Z. Anal. Chem.,* **244,** 302 (1969).
4. AUBEAU, R., CHAMPEIX, L. and REISS, J., *J. Chromatog.,* **16,** 7 (1964).
5. DUSWALT, A. A. and BRANDT, W. W., *Anal. Chem.,* **32,** 272 (1960).
6. KYRYACOS, G. and BOORD, C. E., *Abstracts* 131st *Meeting,* Am. Chem. Soc. (1957).
7. SUNDBERG, O. E. and MARESH, C., *Anal. Chem.,* **32,** 274 (1960).
8. GOLDUP, A. and WESTAWAY, M. T., *Anal. Chem.,* **38,** 1657 (1966).
9. KNIGHT, H. S. and WEISS, F. T., *Anal. Chem.,* **34,** 749 (1962).
10. KUNG, J. T., WHITNEY, J. E. and CAVAGNOL, J. C., *Anal. Chem.,* **33,** 1505 (1961).
11. KAISER, R., *Chromatographia,* **2,** 453 (1969).
12. BROCHE, H. and SCHEER, W., *Brennstoff-Chem.,* **13,** 281 (1932).
13. GELEZUNAS, V. L., *Anal. Chem.,* **41,** 1400 (1969).
14. BAKER, B. B. Jr. and MacNEVIN, W. M., *Anal. Chem.,* **22,** 364 (1950).
15. BEREZKIN, V. G., MYSAK, A. Y. E. and POLAK, L. S., *Neftekhimiya,* **4,** 156 (1964).

16. MUSHA, S. and NISHIMURA, T., *Japan Analyst,* **14,** 803 (1965).
17. ERLEY, D. S., *Anal. Chem.,* **29,** 1564 (1957).
18. CRITCHFIELD, F. E. and BISHOP, E. T., *Anal. Chem.,* **33,** 1034 (1961).
19. HAGER, M. and BAKER, G., *Proc. Montana Acad. Sci.,* **22,** 1962 (1963).
20. MARTIN, J. H. and KNEVEL, A. M., *J. Pharm. Sci.,* **54,** 1464 (1965).
21. CARLSTROM, A. A., SPENCER, C. F. and JOHNSON, J. F., *Anal. Chem.,* **32,** 1056 (1960).
22. SMITH, B., *Acta Chem. Scand.,* **13,** 480 (1959).
23. OKUBO, N., MASHIMO, S., WATANABE, T. and JONO, W., *Japan Analyst,* **15,** 949 (1966).
24. BERESNEV, A. N., FOROV, V. B., MIRZAYANOV, V. S. and MATSNEVA, N. V., *Zh. Analit. Khim.,* **24,** 280 (1969).
25. BURKE, D. E., WILLIAMS, G. C. and PLANK, C. A., *Anal. Chem.,* **39,** 544 (1967).
26. HOLLIS, O. L. and HAYES, W. V., *J. Gas Chromatog.,* **4,** 235 (1966).
27. COTTON, T. M., BALLSCHMITTER, K. and KATZ, J. J., *J. Chromatog. Sci.,* **8,** 546 (1970).
28. HOGAN, J. M., ENGEL, R. A. and STEVENSON, H. F., *Anal. Chem.,* **42,** 249 (1970).
29. HACHENBERG, H., unpublished work.
30. LEONHARDT, J., *Chem. Tech. (Berlin),* **20,** 428 (1968).
31. MYERS, H. S., *Gas,* Los Angeles, **44,** 60 (1968).

2.4. SULPHUR COMPOUNDS

Sulphur compounds are among those trace materials which are particularly undesirable in manufacturing processes since they cause corrosion and poison catalysts. Thus, for example, the sulphur content of petroleum fractions and of petroleum itself is of considerable interest in thermal cracking processes for the manufacture of unsaturated hydrocarbons. There is also the need to monitor traces of sulphur in fuels for motor vehicles and aircraft, where corrosion must be avoided. It is also necessary to analyse gaseous sulphur compounds in the atmosphere.

This type of analytical monitoring must be rapid, sensitive and specific. The well known classical methods of determining the total sulphur concentration in gases and liquids, based on reduction to the sulphide ion or oxidation to the sulphate ion, are employed to a decreasing extent because of the requirements just stated. Moreover, several of these methods, such as Wickbold's oxidative technique, require cumbersome safety precautions owing to the possibility of explosion. In addition, large samples are required to carry out trace analyses at the ppm level, and it is more or less impossible to operate these classical chemical procedures continuously or to automate them.

However, with very few exceptions, the main disadvantage of these classical methods is that only the total concentration of sulphur can be measured, and individual components cannot be determined separately. The exceptions include analysis of hydrogen sulphide which can be specifically determined by the very sensitive molybdenum blue photometric method, or with gas indicator tubes based on the reaction with heavy metal salts.[1] The fluorescence method is another sensitive and selective technique.[2] There have been many attempts to develop gas chromatographic methods for determination of traces of sulphur compounds since the technique is rapid, compared with the classical methods, and with suitable detectors it is specific, very sensitive and also safe. It needs only very small samples, can be automated, and also enables individual constituents to be separated and determined.

ANALYSES USING NON-SPECIFIC DETECTORS

There are no specific detectors available for determination of sulphur compounds so that appropriate methods of separating traces of them from the other principal impurities have to be developed.

Generally it is difficult to achieve these separations, especially in hydro-carbon mixtures, but it is relatively simple in some cases, *e.g.* the determination of odorants. Thus, tetrahydrothiophen (THT) may easily be determined in town gas with an FID since its boiling point of 123°C is well above the boiling points of the hydrocarbons present.[1] Figure 114 shows a typical chromato-gram obtained with a 10 m polypropylene glycol column.

Karchmer has separated eleven sulphur compounds having boiling points in the range 58 to 126°C with β,β'-iminodipropionitrile as stationary phase at 84°C.[3] Gas chromatographic separation of hydrogen sulphide, methyl-, ethyl- and propylmercaptan, dimethyl- and diethylsulphide and thiophen

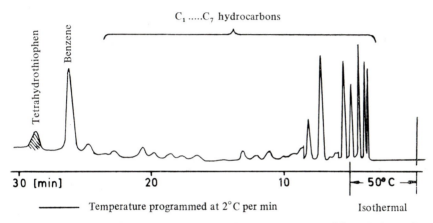

Fig. 114   Determination of tetrahydrothiophen in town gas with an FID and a 10 m polypropylene glycol column, temperature programmed from 50–150°C at 2 deg C/min (from W. Ripperger, *Gas-Wasserfach*, **109**, 375 (1968)).

from alcohols, and sulphur compounds from light hydrocarbons, and also of benzene from thiophen may be achieved with tricresyl phosphate. The limit of detection was reported to be 2×10^{-8} mole for ethylmercaptan.[4] Twelve mercaptans and eleven sulphides boiling in the range 35 to 220°C were separated by Sullivan *et al.* on a 2·1 m squalane column, temperature programmed from 20 to 150°C in 28 minutes.[5] The separation of C_4 and C_5 mercaptans has been reported by Sunner *et al.* and also by Liberti and Cartoni.[6,7] Spencer *et al.* have analysed odorants (mercaptans and sulphides) in natural gas on dinonyl phthalate at 50°C.[8] Thiols and sulphides may also be separated on benzyldiphenyl, tritolyl phosphate and silicone oil.[4,9,10] Schols has used N,N'-di-n-butyl acetamide as stationary phase and a ther-mistor detector to determine COS in the range 25–1000 ppm in natural gas.[11] Adequate resolution of the component from the hydrocarbons present was obtained in an analysis time of 6 minutes. Hachenberg has determined traces of carbonyl sulphide in water gas on a 2 m Porapak R column at 22°C.[12]

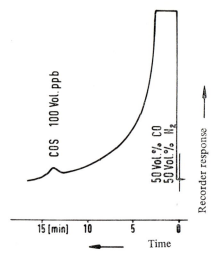

Fig. 115   Determination of traces of COS in water gas using a helium detector and a Porapak R column.

With a helium detector, 50 ppb COS could be measured in a 1 ml sample. This analysis is illustrated in Fig. 115.

Analysis of traces of COS in CO_2 is important in the beverages industry, and Hall has described a method which employed a silica gel column and a β-radiation detector.[13] The limit of detection for COS in CO_2 was 0·3 ppm and for the other sulphur compounds which might be present, *i.e.* H_2S and

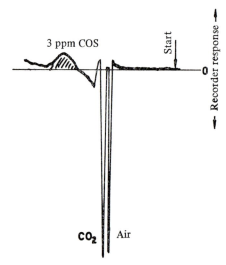

Fig. 116   Determination of COS in CO_2 on a 1·2 m silica gel column at 25°C (from H. L. Hall, *Anal. Chem.*, **34**, 63 (1962)).

CS_2 it was about 0·8 vol ppm. Figure 116 illustrates the determination of 3 ppm COS in CO_2 in which air and CO_2 give negative signals.

Similar analyses of fuel gases for traces of hydrogen sulphide, carbonyl sulphide and organo-sulphur compounds have been carried out by Reinhardt *et al.* using an argon detector having a linear response from 10^{-8} to 10^{-6} g and which gave a limit of detection of 1×10^{-8} g for each component.[14] Dioctyl phthalate and squalane were used as the stationary phases. Table 5 lists the R_F values of some sulphur compounds and hydrocarbons on the squalane column at 10°C.

TABLE 5   **Relative retention times of organo-sulphur compounds and hydrocarbons**

(from Table 1 in M. Reinhardt *et al., Chem. Tech. (Berlin),* **19**, 42 (1967))

Substance	Relative retention times (n-pentane = 1·000, 5 m column, 10% squalane on Porolith 0·3–0·4 mm, Temperature = 10°C)
HCN	0·029
H_2S	0·030
Carbonyl sulphide	0·073
Methylmercaptan	0·269
Isopentane	0·709
Ethylmercaptan	0·802
n-pentane	1·000
Methyl thioether	1·073
Isopropylmercaptan	1·603
Carbon disulphide	1·820

Figure 117 shows the analysis of COS and Fig. 118 the determination of CS_2 in hydrocarbon mixtures.

The investigations of Reinhardt *et al.* showed that the argon detector used was more sensitive to some sulphur compounds than to hydrocarbons. Thus, the response to methyl thioether and hydrogen sulphide was respectively 1·4 and 2 times that for n-pentane. Moreover, it can be seen from Table 6 that for an FID there is definitely no correlation between the responses of the detector to these compounds and their C-numbers.[15]

It has also been found that hydrogen sulphide gives only 1% of the response of methane.[16] The results of Feldstein *et al.* are probably accounted for by the sulphur compounds forming decomposition products in the flame of the FID, which corrode or coat the jet, and this will be observed as a diminution in sensitivity. Feldstein *et al.* also examined the separation of organic sulphur compounds from hydrocarbons on various polar and non-polar stationary phases; Versamid 930, silicone-Triton, Flexol, tritolyl phosphate, Ucon,

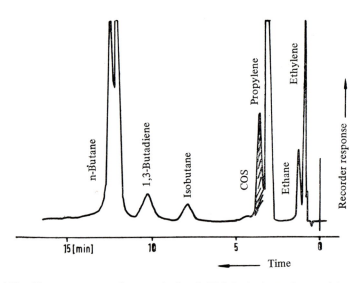

Fig. 117   Chromatogram of an analysis of COS in hydrocarbons with squalane as stationary phase at 14·5°C (from M. Reinhardt *et al., Chem. Tech. (Berlin)*, **19,** 43 (1967)).

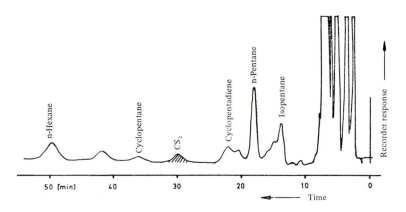

Fig. 118   Chromatogram of an analysis of CS_2 in hydrocarbons with squalane as stationary phase at 40°C (from M. Reinhardt *et al., Chem. Tech. (Berlin)*, **19,** 43 (1967)).

TABLE 6 Relative response of the FID to hydrocarbons and organic sulphur compounds having different C-numbers (from M. Feldstein *et al.*, *J. Air Pollution Control Assoc.*, **15**, 215 (1965))

Substance	Response	C-number
CH_4	1	1
C_2H_6	2	2
C_2H_4	2	2
C_2H_2	2	2
C_3H_8	3	3
C_4H_{10}	4	4
C_6H_6	5·8	6
C_7H_{16}	7·4	7
CH_3OH	0·83	1
C_2H_5OH	2	2
CCl_4	0·64	1
$CHCl_3$	0·85	1
$(CH_3)_3N$	0·58	3
CH_3SH	0·3	1
C_2H_5SH	0·6	2
C_3H_7SH	0·8	3
CH_3SCH_3	0·4	2
CH_3SSCH_3	0·4	2

asphalt, and Apiezon. In no case was it found possible to obtain an efficient resolution of the mercaptans from the hydrocarbons.

These workers also drew attention to a source of error in the trace analysis of sulphur compounds. This results from reaction of the compounds with the solid support, stationary phase, or material from which the column is made. Thus, for example, copper should be avoided, the most suitable materials for the column being Teflon and stainless steel. For the same reason, preparation of ppm standard mixtures of sulphur compounds is rather difficult, and they should always be used immediately after they are prepared. The following measurements (Table 7) made by Feldstein *et al.* show that glass

TABLE 7 Stability of standard ppm mixtures of methylmercaptan in stainless steel and glass containers (from M. Feldstein *et al.*, *J. Air Pollution Control Assoc.*, **15**, 215 (1965))

Time (h)	Decrease in concentration of CH_3SH in ppm			
	stainless steel vessel		glass vessel	
0	5·2	75	5·2	75
24	1·5	65	4·0	73
48	0	53	3·1	74
72	0	47	2·2	70

vessels are more suitable than stainless steel ones for the preparation of these mixtures.

Similar observations have been made by Koppe and Adams in an investigation of gas chromatographic columns suitable for the analysis of gaseous sulphur compounds at concentrations below 1 ppm.[17] They wished to determine traces of hydrogen sulphide, methylmercaptan, dimethyl sulphide, dimethyl disulphide and sulphur dioxide, and examined columns made from aluminium, glass, stainless steel and Teflon together with 12 solid supports and 31 stationary phases. Stainless steel columns packed with 10% Triton X-305 on Chromosorb G/DMCS were recommended, and it was found that aluminium columns retard hydrogen sulphide. The same kind of difficulty was also reported by Stevens *et al.* in the analysis of traces of hydrogen sulphide and sulphur dioxide.[18] It can be concluded from the work of these authors that specific responses cannot be obtained in the determination of traces of sulphur compounds with the FID, thermal conductivity, argon or helium detector in spite of the differences in response towards some other compounds. The success of trace analyses with these detectors therefore depends on choice of column, optimization of operating conditions, and the composition of the sample itself. Before dealing with specific detectors for sulphur compounds, two methods for the selective determination of these materials will be described.[15]

The first method involves *selective concentration*, and this can be achieved with the aid of various salt solutions. For example, hydrogen sulphide can be absorbed by aqueous cadmium sulphate solution, mercaptans by aqueous mercuric cyanide, and thioethers and disulphides by aqueous mercuric chloride. For sample collection, the solutions are in a series of wash bottles and the corresponding sulphur compounds may be liberated with 6N HCl and then analysed gas chromatographically.

Another similar procedure consists of producing two gas chromatograms, one of which is obtained with a short mercuric sulphate column connected in series. No sulphur compounds appear on the latter chromatogram since they are retained in the precolumn. Using this principle, Okita has developed a filter method for the determination of traces of mercaptans and organic sulphides in the region of 1 ppb and less in the atmosphere.[19] Pre-concentration is carried out with mercury salts and regeneration with hydrochloric acid and sodium hydroxide solution, the regenerated sulphur compounds being absorbed in organic solvents. Gas chromatographic analysis of the mercaptans and dimethyl sulphide uses tricresyl phosphate as the stationary phase.

The second method of selective determination depends on *chemical conversion of the sulphur compounds being analysed* into hydrocarbons. This is achieved by hydrogenating the sample with Raney nickel in hot alcoholic solution. The reaction converts thiols, sulphides and disulphides into the corresponding hydrocarbons which are then analysed by gas chromatography. For example, methane is formed from dimethyl sulphide, and methyl n-propyl sulphide gives methane and propane.[20] When the reaction gives gaseous products, a sample is introduced directly into the gas chromatograph, whereas the alcoholic solution is sampled for liquid products.

Another method, which also results in formation of the corresponding hydrocarbons is catalytic desulphurization.[21] Compared with the Raney nickel method, it has the advantage of only requiring a very small sample and it can be performed rapidly. Sulphur compounds may also be converted into sulphur dioxide and carbon dioxide by passing them in a stream of oxygen over a platinum catalyst at 850°C. The two products and also the oxygen can be separated on a dinonyl phthalate column.[22] This method has proved to be satisfactory for determination of sulphoxides, sulphones, sulphides, dimethyl sulphides and thioethers. Sulphates are not quantitatively converted into sulphur dioxide under the above conditions. The limit of detection is 0·05 mg when a thermal conductivity detector is used. Fluorine, chlorine, nitrogen and oxygen compounds do not interfere.

These time-consuming concentration techniques and chemical conversions have now been superseded to some extent, since detectors specific to sulphur are available which enable trace analyses to be performed more easily and rapidly. Moreover, they offer the possibility of automation for continuous plant control.

DETERMINATION OF SULPHUR WITH SPECIFIC DETECTORS

Apart from the mass spectrometer, the microcoulometer, alkali flame ionization detector (AFID) and the more recently developed flame photometric detector (FPD) may be considered for the specific detection of small amounts of sulphur compounds.

DOHRMANN'S MICROCOULOMETER

The microcoulometer, originally developed by Coulson et al. as a gas chromatographic detector for the analysis of sulphur and chlorine compounds, provides oxidative and reductive methods for the determination of sulphur.[23,24]

In the *oxidative method*, the mixture of carrier gas and sample emerging from the gas chromatograph is flushed into a reactor tube heated to 750°C in which it is burned in oxygen. The reactor is made of silica and is packed with silica chips.

The combustion products CO_2, H_2O and SO_2 derived from the sulphur-containing components pass into the titration cell. The SO_2 is oxidized to SO_3 by iodine with the result that the iodine/iodide equilibrium which had been previously established in the cell is disturbed. The iodine consumed is recorded by the measuring electrode as a change in e.m.f. which is amplified in the coulometer. In response to this change in e.m.f., fresh iodine is produced by electrolysis at the generator electrode until the original iodine/iodide equilibrium is restored. The current used to produce the iodine is usually displayed on a recorder as a differential chromatogram. The peak area corresponds to the quantity of electricity required in the oxidation of the SO_2.

The following reactions take place in the iodine cell:

Titration:

$$2I° + SO_2 + H_2O \longrightarrow SO_3 + 2I^- + 2H^+$$

Regeneration:

$$2I^- \longrightarrow 2I° + 2e^-$$

Interference occurs in this oxidative method if combustion is incomplete. In addition, the presence of halogen and nitrogen compounds has a detrimental effect on the titration of sulphur dioxide with iodine. This interference is caused by oxidation of these compounds liberating iodine from the potassium iodide. Consequently the concentration of iodine in the electrolyte is increased, and a negative peak obtained. According to Killer this effect is noticeable even at a nitrogen or chlorine concentration in the sample of 0·1%. The effect may be overcome by adding sodium azide which rapidly reacts with chlorine and oxides of nitrogen, but only slowly with iodine. Bromine causes considerably greater interference in this analysis but it cannot be removed with sodium azide since the reaction proceeds too slowly. This fact precludes the use of the microcoulometer for analysis of sulphur in leaded petrols since these generally contain the additive dibromoethane.[25] A further difficulty to consider arises in the analysis of compounds having high boiling points since they can form carbon deposits in the combustion tube on which the products of pyrolysis can be absorbed.

Nevertheless, the reproducibility of the oxidative method is generally good. For example, Drushel has reported a relative deviation of ± 2–4% for concentrations exceeding 300 ppm and an absolute error of $\pm 0·3$–0·6 ppm in the range 1–10 ppm.[26]

A linear relationship exists between peak area and sulphur content up to about 200 ng of sulphur. If the sample contains larger amounts of sulphur, flat peaks are often obtained. In this case it is advisable to dilute the sample with a sulphur-free hydrocarbon before analysis.

MICROCOULOMETRIC DETERMINATION OF SULPHUR BY REDUCTIVE PYROLYSIS produces hydrogen sulphide which is titrated with electrolytically formed silver as follows:

Titration:

$$2Ag^+ + S^{2-} \longrightarrow Ag_2S$$

Regeneration:

$$Ag° \longrightarrow Ag^+ + e^-$$

The reduction method is not affected by halogens, but nitrogen compounds do interfere.

By careful control of the operating conditions, and taking the above sources of interference into account, acceptable results may be obtained from both the oxidative and reductive procedures. Ripperger has determined traces of sulphur compounds in natural gas by the oxidation method.[1] This analysis is illustrated in Fig. 119, in which the upper chromatogram was obtained with an FID, whilst the lower one represents selective recording of traces of sulphur-containing components with a coulometer.

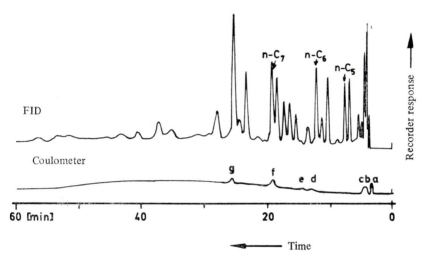

Fig. 119   Specific determination of traces of sulphur compounds with a coulometer and a 10 m polypropylene glycol column, temperature programmed from 40 to 160°C (from W. Ripperger, *Gas-Wasserfach*, **109**, 376 (1968)). *a* = hydrogen sulphide, *b* = dimethyl sulphide, *c* = methylmercaptan, *d* = propylmercaptan, *e* = unknown, *f* = propylmercaptan-1, *g* = butylmercaptan-2.

Fredericks and Harlow have determined traces of mercaptans in the range 0·8 to 150 vol ppm in natural gas containing large amounts (1·5–15%) of hydrogen sulphide.[27] The precision was $\pm 2\%$ at 50 ppm. Since the high concentration of hydrogen sulphide overloaded the detector, the compound was removed through a by-pass together with the low-boiling hydrocarbons. Martin and Grant have used a coulometer down to 5 ppm in their investigation of the distribution of sulphur compounds in different petroleum fractions.[28] Studies of the extent of sulphur dioxide formation as a function of combustion temperature in the oxidation method have shown that it is not 100%. The maximum for this conversion is at 650–700°C as the experimental results in Table 8 demonstrate.

The yield of sulphur dioxide again decreases above 700°C. The age of the silica packing in the combustion tube also affects the formation of sulphur dioxide. By using standard mixtures containing known total concentrations of sulphur, it was shown that the error in the gas chromatographic results is only 3 %.

The limit of detection of the coulometric method is generally 1 ppm and consequently between that for an FID and a good thermal conductivity detector. A limit of 2 ppm has been reported for the determination of sulphur in gases and low-boiling liquids.[25] Disadvantages of the coulometer compared with other detectors are its relatively large time constant, and the overloading which occurs at high sulphur concentrations (broad peaks and tailing in the measuring cell).

TABLE 8 **Formation of sulphur dioxide as a function of temperature** (from R. L. Martin and J. A. Grant, *Anal. Chem.*, **37**, 644 (1965))

Combustion temperature °C	Conversion to SO_2 %
550	70
600	80
650	91
700	93
750	89
850	74
950	63

A variation of this method consists of using bromine instead of iodine to titrate the sulphur dioxide.[29] Adams *et al.* have converted the iodine cell in Dohrmann's coulometer into a bromine cell by replacing the iodine in the reference cell with elementary bromine.[30] A mercuric-mercurous bromide paste was used for the reference electrode. The polarity has to be changed to give a positive output signal. The tailing which commonly occurs with the I_2 cell is significantly reduced by this technique as shown in Fig. 120.

The lower detectable limits reported for this method are H_2S: 10 ppb, SO_2: 50 ppb, CH_3SH: 10 ppb, CH_3SCH_3: 50 ppb and CH_3SSCH_3: 5 ppb.

The coulometric methods of analysis described above are specific and sensitive, but they cannot be incorporated in equipment used for process control, particularly of flowing gases, or for atmospheric monitoring since the apparatus is too expensive and very fragile owing to the silica and glass components. The use of coulometers is therefore generally confined to laboratories.

The only detectors which can be considered for plant and process control are those which can easily be incorporated within the gas chromatographic unit. For a number of years process gas chromatographs have therefore been equipped with the FID.

The alkali flame ionization detector (AFID) (*see* 1.31) is specific for sulphur compounds under certain experimental conditions, but has not yet been applied to the trace analysis of this class of compound.[31] It should be noted that some detectors of this type require to be operated for a certain time before they attain a constant sensitivity.[32]

The flame photometric detector (FPD) is recommended for the very low concentration regions (less than 1 ppm) since it can still give quantitative results at a level where the coulometer has become unreliable. The principle of the FPD depends on the photometric detection of the flame emission

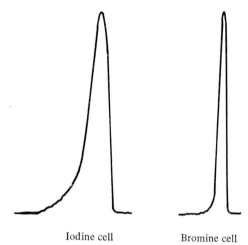

Iodine cell   Bromine cell

Fig. 120 Comparison of an H_2S trace peak from a bromine cell with that from an iodine cell (from D. F. Adams *et al., Anal. Chem.,* **38,** 1094 (1966)).

from the S_2 species in a hydrogen–air flame.[33,34] Optical filtering is confined to the very narrow region of 394 ± 5 nm. The emission is selectively sensed by a photomultiplier, amplified, and the signal registered on a recorder in the usual way. This system makes it possible to specifically differentiate sulphur compounds from the other constituents in a sample. The operating principle of this detector is illustrated in Fig. 121.

Brody and Chaney were the first to examine the performance of this detector in combination with gas chromatography, and they recommended the following operating conditions.[35]

Detector: FPD
Temperature: 235°C
Temperature of the flame photometer burner housing: 100°C
Nitrogen flow: 160 ml min^{-1} (carrier gas)
Oxygen flow: 40 ml min^{-1}
Hydrogen flow: 200 ml min^{-1}
Base line noise: $1\cdot5 \times 10^{-10}$ A

Under these conditions it was found that the response is approximately proportional to the square of the concentration over the range 1–1000 ppm. The authors reported a limit of detection of 0·6 ppm for malathion and parathion using the sulphur filter at 394 nm. Mizany has studied the relationship between the response of different sulphur compounds (diethyl sulphide, diethyl disulphide, diethylsulphone, diethyl sulphite, diethyl sulphate) and the oxidation state of the sulphur in the molecule.[36] The response of the FPD depends on the emission from the S_2 species formed in the flame. The yield of this species is related to the oxygen–hydrogen ratio, and consequently to the flame temperature. Mizany found that the highest sensitivity is obtained with an oxygen–hydrogen ratio of 0·4–0·5.

Goode has analysed traces of sulphur compounds in natural gas from the North Sea.[37] The compounds were identified by measuring their retention times on 3 m tricresyl phosphate and 3·6 m squalane columns. It is important

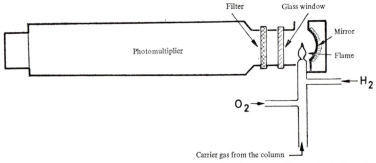

Fig. 121   Construction of the flame photometric detector (from S. S. Brody and J. E. Chaney, *J. Gas Chromatog.,* **4,** 42 (1966)).

that the column should be made of aluminium, and the solid support must be silanized, since otherwise severe tailing will seriously affect the quantitative results. It was recommended that the samples of sulphur-containing gases should be collected in aluminium cylinders fitted with aluminium valves. It is necessary to clean these containers with carbon tetrachloride and dry them with compressed air before collecting the sample. Sampling should be carried out by repeated evacuation and filling of the vessel. This procedure almost entirely avoids changes in the sample composition resulting from atmospheric oxidation of the mercaptans to disulphides. Since disulphides are not present in natural gas, Goode recommended that oxidation with oxygen or iodine should be carried out prior to the analyses. The disulphides formed are then measured instead of the mercaptans originally present, since the former are more easily resolved gas chromatographically.[38] After allowing for the signal-to-noise ratio, the limits of detection are found to depend on the retention time. In a 1 ml sample, C_2–C_4 sulphides can be detected in concentrations which are equivalent to 0·02 ppm by weight of sulphur. Even

0·005 ppm by weight can be measured in a 5 ml sample. At very low concentrations, the peak height is proportional to concentration, but this linearity of the FPD no longer holds at higher concentrations. Bowman and Beroza have shown that over the concentration range 1–1000 ppm of sulphur compounds the peak height produced by this detector is proportional to the square of the concentration.[39] They employed an FPD having two photomultipliers, one of which was fitted with a 526 nm interference filter for phosphorus, and the other had a 394 nm filter to detect the S_2 species. The two signals permitted complete differentiation of phosphorus and sulphur in molecules which contained both.

The artificial additives (odorants) in natural gas may be elegantly and simply analysed with the FPD. Thus, methyl-, ethyl-, isopropyl-, and n-propylmercaptan may be determined in ppb amounts on PTFE treated with 0·5% phosphoric acid, and temperature programmed from 50 to 100°C in 10 minutes. In addition to its high sensitivity and specificity, the FPD has the great advantage of a stable base line during analysis and reproducibility of its output. These characteristics result in the detector being increasingly used in practice, particularly for the continuous monitoring of traces of sulphur-containing components in gaseous process streams. Hachenberg employed this analytical technique to determine ppb amounts of COS and H_2S in water gas.[12] The procedure included direct sampling from the process stream and was carried out automatically over the range 0·1–0·8 vol ppm. A laboratory gas chromatograph equipped with an automatic sample injector was used under the following conditions:

Instrument: Varian-MAT, model 1400
Detector: Flame photometer, Techmation, model 100-AT
Detector temperature: 140°C
Combustible gases: hydrogen: 150 ml min^{-1}
                  oxygen: 20 ml min^{-1}
                  air: 50 ml min^{-1}
Column: glass, 4 mm internal diameter, length 2 m, packed with 80/100 mesh Porapak R
Column temperature: 80°C
Sample size: 24 ml with an automatic injector, Carlo Erba No. 65322
Recorder: Philips, PM 8000, range 1 mV.

Figure 122 shows the continuous recording of traces of COS and H_2S in the region below 1 ppm. There is a linear relation between the peak height and concentration up to 1 ppm.

Continuous sampling was also employed by Stevens and O'Keeffe in the analysis of ppm amounts of hydrogen sulphide, sulphur dioxide, and methyl- and ethylmercaptan in air.[41] They confirmed the above linearity in the range 5 ppb up to about 1 ppm (Figs. 123 and 124).

If it is not required to determine the individual sulphur compounds, then the total sulphur concentration may be continuously measured by by-passing the column. This is illustrated in Fig. 125 where the total amount of sulphur in the atmosphere is recorded as sulphur dioxide.

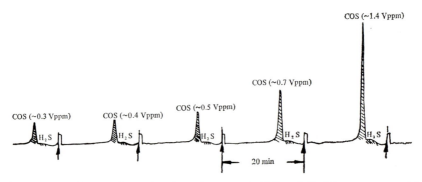

Fig. 122   Continuous determination of traces of COS and H_2S in water gas with the FPD.

When the FPD is used to measure sulphur, either as the total amount of the element, or for the specific analysis of individual sulphur compounds, quenching by the major component present must be taken into account, particularly when the FPD is known to show a high response to it. Perry and Carter studied this quenching effect, which can severely diminish the response to sulphur compounds, and stressed the point that special attention should be paid to it in trace analysis.[42] For example, the response to 10 ppm thiophen is reduced

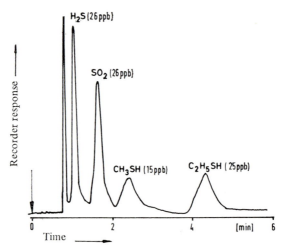

Fig. 123   Determination of traces of sulphur compounds in air with the FPD (from R. K. Stevens and A. E. O'Keeffe, *Anal. Chem., 42*, 143A (1970)).

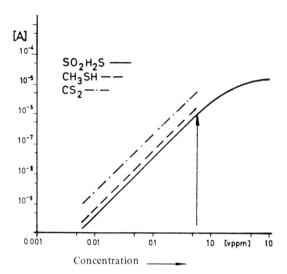

Fig. 124   Linear relationship between the response of the FPD and the concentration of sulphur compounds in the ppb region (from R. K. Stevens and A. E. O'Keeffe, *Anal. Chem.*, **42**, 143A (1970)).

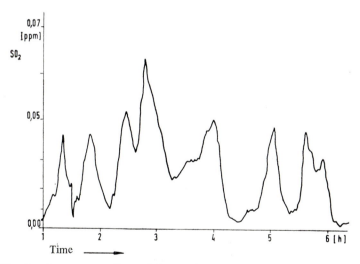

Fig. 125   Continuous recording of the total concentration of sulphur in air with the FPD (from R. K. Stevens and A. E. O'Keeffe, *Anal. Chem.*, **42**, 143A (1970)).

by 80% when 1% benzene is present. For accurate quantitative trace analysis it is therefore necessary to ensure complete resolution of the sulphur-containing components being analysed from the other compounds present. In the determination of total sulphur content, *e.g.* in fuel oils and petrols, the hydrocarbons are oxidized to CO_2, which produces only a very slight quenching effect.[43] A detector based on a combination of this quenching effect and measurement of the chemiluminescence has been proposed by Crider and Slater.[44] Its sensitivity towards sulphur and halogen compounds should be comparable with that of a flame ionization detector. On the other hand, the sensitivity towards hydrocarbons and oxygen-containing compounds is similar to that of a thermal conductivity detector.

References
1. RIPPERGER, W., *Gas-Wasserfach,* **109,** 377 (1968).
2. AXELROD, H. D., CARY, J. H., BONELLI, J. E. and LODGE, G. P. Jr., *Anal. Chem.,* **41,** 1856 (1969).
3. KARCHMER, J. H., *Anal. Chem.,* **31,** 1377 (1959).
4. RYCE, S. A. and BRYCE, W. A., *Anal. Chem.,* **29,** 925 (1957).
5. SULLIVAN, J. H., WALSH, J. T. and MERRITT, C. Jr., *Anal. Chem.,* **31,** 1826 (1959).
6. SUNNER, S., KARRMAN, K. J. and SUNDEN, V., *Mikrochim. Acta,* 1144 (1956).
7. LIBERTI, A. and CARTONI, G. P., *Chim. Ind.* (Milan), **39,** 821 (1957).
8. SPENCER, C. F., BAUMANN, F. and JOHNSON, J. F., *Anal. Chem.,* **30,** 1473 (1958).
9. DESTY, D. H. and WHYMAN, B. H. F., *Anal. Chem.,* **29,** 320 (1957).
10. COLEMAN, H. J., THOMPSON, C. J., WARD, C. C. and RALL, H. T., *Anal. Chem.,* **30,** 1592 (1958).
11. SCHOLS, J. A., *Anal. Chem.,* **33,** 359 (1961).
12. HACHENBERG, H., *Erdöl Kohle,* **24,** 630 (1971).
13. HALL, H. L., *Anal. Chem.,* **34,** 61 (1962).
14. REINHARDT, M., KOCH, H. and OTTO R., *Chem. Tech.* (Berlin), **19,** 42 (1967).
15. FELDSTEIN, M., LEVAGGI, D. A. and BALESTRIERI, S., *J. Air Pollution Control Assoc.,* **15,** 215 (1965).
16. DOUGLAL, D. M. and SCHAEFER, B. A., *J. Chromatog. Sci.,* **7,** 433 (1969).
17. KOPPE, R. K. and ADAMS, D. F., *Environm. Sci. Technol.,* **1,** 479 (1967).
18. STEVENS, R. K., O'KEEFFE, A. E., MULIK, J. D. and KROST, K. J., 157*th Meeting,* Amer. Chem. Soc. Div. Anal. Chem., Minneapolis, April 14–18 (1969) Papers, ANAL 56.
19. OKITA, T., *Atmospheric Environ.,* **4,** 93 (1970).
20. FRANC, J., DVORACEK, J. and KOLOUSKOVA, V., *Mikrochim. Acta,* 1 (1965).
21. THOMPSON, C. J., COLEMAN, H. J., WARD, C. C. and HALL, H. T., Atlantic City Meeting 1959, 13–18 September.
22. BEUERMAN, D. R. and MELOAN, C. E., *Anal. Chem.,* **34,** 319 (1962).
23. COULSON, D. M., CAVANAGH, L. A., De VRIES, J. E. and WALTER, B., *J. Agric. Food Chem.,* **8,** 399 (1960).
24. WALLACE, L. D., KOHLENBERGER, D. W., JOYCE, R. J., MOORE, R. T., RIDDLE, M. E. and McNULTY, J. A., *Anal. Chem.,* **42,** 387 (1970).

25. KILLER, F. C. A., *Erdöl Kohle,* **23,** 655 (1970).
26. DRUSHEL, H. V., *Erdöl Kohle,* **22,** 355 (1969).
27. FREDERICKS, E. M. and HARLOW, G. A., *Anal. Chem.,* **36,** 263 (1964).
28. MARTIN, R. L. and GRANT, J. A., *Anal. Chem.,* **37,** 644 (1965).
29. KLAAS, P. J., *Anal. Chem.,* **33,** 1851 (1961).
30. ADAMS, D. F., JENSEN, G. A., STEADMAN, J. P., KOPPE, R. K. and ROBERTSON, T. J., *Anal. Chem.,* **38,** 1094 (1966).
31. DRESSLER, M. and JANÁK, J., *J. Chromatog. Sci.,* **7,** 451 (1969).
32. HACHENBERG, H. and GUTBERLET, J., *Brennstoff-Chem.,* **49,** 246 (1968).
33. DRAEGERWERK, H. and DRAEGER, B., West German Patent 1,133,918 July 26 (1962).
34. CRIDER, W. L., *Anal. Chem.,* **37,** 1770 (1965).
35. BRODY, S. S. and CHANEY, J. E., *J. Gas Chromatog.,* **4,** 42 (1966).
36. MIZANY, A. I., *J. Chromatog. Sci.,* **8,** 151 (1970).
37. GOODE, K. A., *J. Inst. Petrol.,* **56,** 33 (1970).
38. SPOREK, K. F. and DANYI, M. D., *Anal. Chem.,* **35,** 956 (1963).
39. BOWMAN, M. C. and BEROZA, M., *Anal. Chem.,* **40,** 1448 (1968).
40. HARTMANN, C. H. and MITCHELL, B., *Research Notes,* Varian Aerograph 11/70.
41. STEVENS, R. K. and O'KEEFFE, A. E., *Anal. Chem.,* **42,** 143A (1970).
42. PERRY, S. G. and CARTER, F. W. G., 8th Int. Symp. on GC, Dublin 1970, Preprints (Eds N. STOCH and S. G. PERRY), Paper 22.
43. RUPPRECHT, W. E. and PHILLIPS, T. R., *Anal. Chim. Acta,* **47,** 439 (1969).
44. CRIDER, W. L. and SLATER, R. W. Jr., *Anal. Chem.,* **41,** 531 (1969).

2.5. SYNTHETIC POLYMERS

There is a wide variety of applications of gas chromatographic trace analysis in the field of synthetic polymer production and research since gases, liquids and solids have to be analysed.[1] For example, it is necessary to determine the impurities in monomers and dispersing agents, and the purity of gases used in polymerization reactions. Also, the polymer itself makes exacting demands in the analytical determination of residual monomers, solvents, plasticizers, anti-oxidants and other additives. It is particularly important to determine these constituents when the polymers concerned are to be used for the packaging of foods, or in the form of aqueous dispersions for interior paints. In polymer research, measurement of residual monomer concentration is important for kinetic studies, especially when they involve copolymerizations. Characterization of polymers by pyrolytic gas chromatography or other thermal and chemical degradation processes also frequently involves trace analysis.

2.51. Purity of Monomers

Knowledge of the trace impurities in monomers is of crucial importance for two reasons connected with the manufacture of polymers by poly-addition or poly-condensation reactions. Certain impurities lead to increased catalyst consumption in chain temination or branching reactions. Other impurities, which do not take part in the polymerization reactions, become concentrated in the polymers formed, and as a result of subsequent reactions can produce coloration, flaws and other undesirable features, or induce secondary condensation reactions in condensation polymers.

ETHYLENE

Ethylene is one of the most important raw materials for the heavy organic chemicals industry. It is at present manufactured by pyrolysis of gaseous hydrocarbons, petroleum or petroleum fractions, and the ethylene obtained can contain different impurities depending on which manufacturing process has been used.[2] For example, in high temperature pyrolysis at 2000–3000°C, in addition to acetylene, considerably more higher acetylenes, NO, NO_2, and N_2O are formed than in the intermediate temperature pyrolysis (steam cracking) at 800–900°C. In addition, the sulphur content of the raw material determines the amounts of H_2S, COS and other undesirable sulphur compounds formed. Depending on the manufacturing process and raw material quality, the following trace constituents may be present in the ethylene (Table 9).

TABLE 9 Compounds which have to be analysed in the different ethylene-manufacturing processes and in the products derived from ethylene

Compound	Boiling point at 760 mm (°C)
Hydrogen	−252·8
Nitrogen	−195·9
Carbon monoxide	−191·6
Oxygen	−183·0
Methane	−161·5
Nitric oxide	−151·8
Ethylene	−103·7
Ethane	−88·6
Nitrous oxide	−88·7
Acetylene	−84·0
Carbon dioxide	−78·5
Hydrogen sulphide	−60·4
Carbonyl sulphide	−50·0
Propylene	−47·7
Propane	−42·1
Propadiene (allene)	−34·5
Methylacetylene	−23·2
Isobutane	−11·7
Isobutylene	−6·9
Butene-1	−6·3
1,3-Butadiene	−4·4
n-Butane	−0·5
Butene-2 (*trans*)	+0·9
Butene-2 (*cis*)	+3·7
Vinylacetylene	+5·1
Ethylacetylene	+8·1
Diacetylene	+10·3
1,2-Butadiene	+10·9
Dimethylacetylene	+27
Nitrogen dioxide	—

Table 9 is not claimed to be complete but it does indicate the large number of compounds which have to be considered in the trace analysis of process streams in an ethylene plant, and in the analysis of the end product, the pure ethylene. Methanol and acetone should also be mentioned since they are contained in the de-icing fluid used in the heat exchangers and fractionating columns. If the ethylene is to be used in the manufacture of organic intermediates such as acetaldehyde, acetic acid, ethanol, ethyl chloride, ethylene oxide, etc. then normal gas chromatographic determination of the purity is generally sufficient, since the degree of purity required in the raw material for these processes is not as high as for the production of polyethylene. In the latter case, the purity required can only be monitored with the most modern analytical instruments.

Of the processes presently employed for the manufacture of polyethylene (high, intermediate and low pressure), the low pressure Ziegler process demands the greatest analytical control, since the organo-metallic compounds used as catalysts can interact with all of the possible impurities. Early attempts were therefore made to determine the purity of ethylene. The non-olefinic impurities had already been determined gas chromatographically in 1955 by means of a concentration technique based on Janák's method.[3,4] A limit of detection of 0·01 vol % was achieved for hydrogen, oxygen, nitrogen, methane, CO and ethane in a 50 ml sample.

Traces of impurities may be concentrated in a similar manner in a nitrometer filled with sulphuric acid saturated with silver sulphate.[5] Subsequent gas chromatographic analysis of the fraction collected enables *e.g.* traces of N_2O to be measured down to the ppb level.

An alternative method depends on concentration by means of 5 Å molecular sieve. In this way it is possible to determine 1 ppm each of carbon monoxide, oxygen, nitrogen and methane in a 1 litre sample in $1\frac{1}{2}$ hours.[6] Concentration on activated charcoal enables less than 6 ppm CO and less than 10 ppm CH_4 and acetylene to be measured with a 200 ml sample. The acetylene is desorbed at 105°C.[7] Brenner and Ettre have concentrated trace amounts of acetylene down to 0·1 ppm in a 40 litre sample by means of a polyethylene glycol precolumn installed immediately after the gas sampling unit on the gas chromatograph.[8]

These concentration techniques are very useful in particular cases, *e.g.* for specialized research applications, but are unsuitable for continual monitoring of production processes, owing to the fact that they take so long. In the case of ethylene, complete analytical procedures have been developed recently for the very rapid determination of trace impurities by a direct method. This is essential if incorrect material feeds are to be avoided in the polymerization process. In this connection, development of sensitive detectors is just as important as that of columns having the highest possible selectivity and thermal and chemical stability. In addition, it is highly desirable to be able to determine a large number of trace components on a single column.

Thus, for example, with a thermal conductivity detector and relatively large samples (about 25 ml), trace materials were able to be measured directly at the following concentration levels: 5 mole ppm hydrogen, 4 mole ppm oxygen, 7 mole ppm methane, 60 mole ppm ethane, 50 mole ppm each of propane and propylene, and 100 mole ppm of C_4 hydrocarbons.[9] It was necessary to use tetraisobutylene, 13X molecular sieve and two different grades of silica gel as the column packings.

Traces of hydrogen, carbon monoxide, methane, propane, propylene, acetylene, n-butane and isobutane have been determined in a 20 ml sample by using a combination of an NaOH-coated Al_2O_3 column and a dibutyl phthalate column, the former being temperature programmed from 20 to 130°C. Overloading of the detector was avoided by bleeding the main component together with the carrier gas into the atmosphere just before the detector. After this operation analysis was continued normally, although the ethane present was lost since it was in the tail of the main component.[10] Nitrogen, methane and carbon dioxide can be determined on aromatic polymers where they elute before the main component, ethylene.[11] Propane, propylene, butane and butene may be analysed on a hexadecene-1 column.[12] Gol'bert and Alekseeva have measured less than 0·2% propylene with an error of 10–15%.[13]

Propyne (methylacetylene) and propadiene (allene) have been determined at the ppm level on alumina coated with β,β'-oxydipropionitrile, dimethyl sulphoxide or dimethylsulpholane.[14] The C_1–C_4 hydrocarbons can be analysed with lower detectable limits of 0·3–0·5 ppm on a combination of columns consisting of alumina coated with 8% sodium bicarbonate (1) and di-isodecyl phthalate (2). Column 1 is programmed from 20 to 70°C, whilst column 2 is operated at room temperature.[15] Similarly, dimethylsulpholane, and also a combination of 5 Å molecular sieve and silica gel, enable hydrogen, nitrogen, carbon monoxide, oxygen, methane and C_1–C_4 hydrocarbons to be analysed in the region of 10–15 ppm.[16]

Of all the impurities mentioned so far, the most difficult compounds to analyse are those which can result from the final purification stage in the manufacture of pure ethylene, *i.e.* distillation (*see* Table 9).

Determination of acetylene is of the greatest importance. Acetylene and ethylene can be separated gas chromatographically on many columns, but analysis of the former in the region below 10 ppm is very difficult since its boiling point is higher than that of ethylene. On most of the columns used for the analysis of gases, therefore, acetylene appears in the tail of the ethylene peak. If, because of their base line stability, uncoated adsorption columns are used for this analysis, then the tailing effect is very marked and, for example, in the case of silica gel it prevents analysis of less than 10 ppm (Fig. 126) if, as is usual with industrial chromatographs, the method of 'peak cutting' is not used.[17]

Derst *et al.* have developed a special method for routine operation in the analysis of traces of acetylene.[18] The column packing consisted of Houdry catalyst coated with 11% dimethylsulpholane. Although acetylene is again eluted after ethylene, it is possible to determine as low as 0·5 ppm by using an FID. Down to 0·5 ppm methane and less than 5 ppm propylene and propane can also be determined at the same time (Fig. 127).

Al'perovich has defined the problems involved in continuously monitoring the quality of high purity ethylene, and discussed the requirements for a

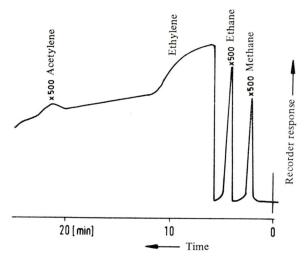

Fig. 126   Determination of 30 ppm acetylene in ethylene on a 3 m silica gel column at 25°C with an FID.

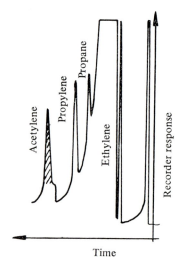

Fig. 127   Routine analysis of acetylene in ethylene with the Houdry catalytic column (length 12 m, Houdry catalyst H 2406 with 11 % by weight dimethylsulpho-lane, column temperature 21°C, 5 ml sample) (from P. Derst *et al., Dechema Monograph.*, **62,** 185 (1968)).

suitable g.c. instrument.[19] He concluded that the reproducibility of the instrument, with regard to sampling from the process stream, can be just as crucial as the carrier gas purity. He carried out the separation of acetylene from ethylene on alumina, and used an 8 ml sample. Hydrogen, methane and ethane may be determined at the same time, the analysis taking 30 minutes. If such relatively large samples are used with thermal conductivity detectors, there is the danger that the filaments will burn out when the main component enters the detector. Therefore, the bridge current is temporarily disconnected before the main peak appears. A typical chromatogram is shown in Fig. 128.

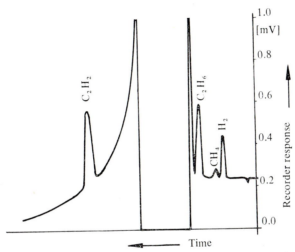

Fig. 128 Determination of traces of acetylene in ethylene on an alumina column with a thermal conductivity detector (from V. Ya. Al'perovich, *Zavodsk. Lab.*, **30**, 1317 (1964)).

If columns with stationary phases coated on an inert solid support are used, then the tailing effect is much less than with the above column packings, but there is interference owing to bleeding of the stationary phase. A cold trap, or low temperature column between the chromatographic column outlet and the detector is then absolutely essential. With a column of HMPA (hexamethylphosphoramide) on Chromosorb, 2·5 ppm C_2H_2 can be determined in 10 minutes in a 5 ml sample, the noise level being 0·02 mV with an FID.[20] Other authors have employed the argon ionization detector for this analysis.[21] The routine measurement of less than 1 ppm acetylene can be performed with dimethylsulpholane on firebrick. Bleeding is reduced by operating the column at 10°C. Rennhak *et al.* have separated acetylene from ethylene with dimethyl sulphoxide on alumina, and obtained a lower detectable limit of 2 ppm of acetylene with an argon ionization detector.[22] A cold trap at a temperature of $-130°C$ was located between the column and detector for the reasons already given.

Those methods in which acetylene elutes before ethylene are always to be preferred. Although the boiling point of acetylene is higher than that of ethylene they can be eluted in the above order on non-polar stationary phases, *e.g.* propylene oligomer.[23] Schwenk *et al.* used decalin at 0°C for this analysis, but owing to the marked bleeding effect this was later replaced by perhydrophenanthrene.[17,24] With this column, it is also possible to analyse all of the

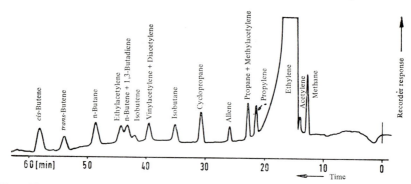

Fig. 129   Determination of traces of acetylene and other hydrocarbons in ethylene on a 20 m perhydrophenanthrene column at 22°C with an FID. Sample size 2 ml.

C_3 and C_4 hydrocarbons at the same time (Fig. 129). The limit of detection of C_2H_2 is 0·1 ppm and between 0·5 and 2 ppm for the other hydrocarbons. Vigdergauz *et al.* have accomplished a similar separation on an 8 m n-heptadecane column.[25] This column is preferable from the point of view of bleeding of the stationary phase, since the boiling point of heptadecane is higher than that of perhydrophenanthrene. The limit of detection for C_2H_2 is

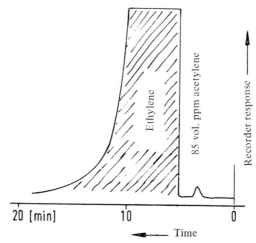

Fig. 130   Analysis of traces of acetylene in ethylene on a 0·6 m carbon molecular sieve column at 150°C (from A. Zlatkis *et al., J. Chromatog. Sci.,* **8,** 417 (1970)).

again in the ppb region. Another column suitable for this analysis is carbon molecular sieve (CMS) since C_2H_2 is eluted before C_2H_4 and even at high temperatures there is no bleeding[26-28] (*see* Fig. 130).

An infrared detector can also be used for the specific determination of acetylene in ethylene and a limit of 5 ppm has been reported for this method.[29]

The principal impurity in high purity ethylene is *ethane*, which does not interfere with chemical processes or polymerizations. Its accurate quantitative measurement is very important for specification of the purity of a sample of ethylene since this is never determined by direct measurement, but by subtracting the concentrations of the impurities from 100%. Hence when ethane is being analysed it should not be in the tail of the ethylene peak owing to the possibility of large errors occurring.

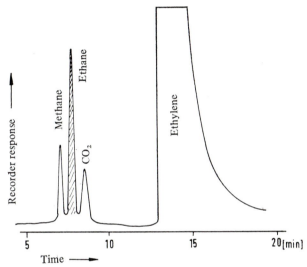

Fig. 131   Determination of ethane in ethylene with *p*-xylyl cyanide–silver nitrate stationary phase at 25°C (9 m column) (from A. Zlatkis *et al., Anal. Chem.,* **36,** 2354 (1964)).

As already mentioned, ethane may be determined on silica gel, where it elutes before ethylene (*see* Fig. 126). Stationary phases which contain silver nitrate are also recommended for this analysis since they selectively retard unsaturated hydrocarbons. A typical analysis using a *p*-xylyl cyanide stationary phase containing silver nitrate is illustrated in Fig. 131.[30]

The gas chromatographic analysis of *traces of carbon dioxide in high purity ethylene* or in C_2–C_3 hydrocarbons generally presents difficulties since the CO_2 coincides with other components on columns normally used for the analysis of gases. For example, CO_2 and acetylene have the same retention time on perhydrophenanthrene and this also applies to CO_2 and propane on acetonyl acetone. Carbon dioxide is therefore usually determined on activated charcoal, since it does not then overlap any other gaseous component. Even if high temperatures are used, it is essential to deactivate the charcoal, *e.g.* by coating with phosphoric acid, since some kinds of the material can irreversibly adsorb relatively large amounts of CO_2. The analysis of 0·5 vol % CO_2 in ethylene on uncoated activated charcoal is compared in Fig. 132 with the same analysis using activated charcoal coated with phosphoric acid.[5]

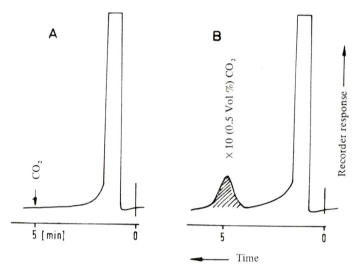

Fig. 132 Determination of traces of CO_2 on uncoated activated charcoal (*A*) and on activated charcoal coated with phosphoric acid (*B*). (1 m column at 22°C.)

Similar problems often arise in the analysis of traces of N_2O on 5 Å molecular sieve. Traces of CO_2 and N_2O could not be measured in gaseous monomers until microporous polymers, derived from styrene and divinylbenzene, became available.[31] Figure 133 illustrates the simultaneous analysis of traces of N_2O and CO_2. It has been found advantageous to use a combined Porapak Q–Porapak T column, since on Porapak Q alone the peak for trace amounts of N_2O can be masked by the ethylene peak.[5]

The helium detector permits measurement of both of these components down to the ppb level. There are no simple methods of determining *traces of carbonyl sulphide and hydrogen sulphide* in high purity ethylene, since both compounds are eluted after ethylene on all types of Porapak.[31] Figure 134 shows this analysis performed on Porapak Q with a thermal conductivity detector.[32]

Since both COS and H_2S generally have to be analysed at levels below 1 ppm, measurements made in the tail of the ethylene peak will be unsatisfactory, even if a helium detector is employed. It is therefore necessary to

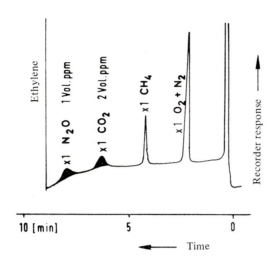

Fig. 133 Determination of traces of N_2O and CO_2 in ethylene on a combined Porapak Q–Porapak T column at 22°C (column lengths 2 m each).

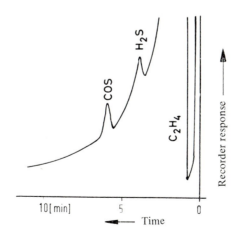

Fig. 134. Determination of traces of COS and H_2S in high purity ethylene with a thermal conductivity detector and a 2 m Porapak Q column at 22°C.

utilize detectors which respond to specific classes of compounds, such as the coulometer or flame photometric detector. Figure 135 shows the selective detection of these two compounds in 99·99 % ethylene with an FPD.[32]

Quantitative calibration of small amounts of H_2S is extremely difficult (*see* 2.4). Moreover, collection of samples of this compound should always be carried out directly from the process stream, and under no circumstances should it be discontinuous.

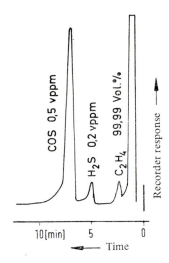

Fig. 135   Determination of traces of COS and H_2S in 99·99 % pure ethylene on a
2 m Porapak Q column at 22°C with an FPD.

As in the case of hydrogen sulphide, reproducible analysis of traces of methanol, acetone and water is only possible if sampling is carried out continuously through a narrow sampling tube connected directly to the gas chromatograph. These compounds can appear intermittently in the ethylene when de-icing operations are carried out during the ethylene purification. Continuous monitoring is therefore necessary during the operation of the process.

The analysis of water is not usually carried out by gas chromatography since a more satisfactory method is with continuously indicating moisture meters, but gas chromatography must be employed to determine the methanol and acetone. This involves similar problems of separation as in the case of COS and H_2S. Methanol has to be measured in the tail of the ethylene peak, and this can only be carried out reliably down to 20 vol ppm, and is very difficult at concentrations below 10 vol ppm. The limit for detection of methanol in ethylene with laboratory gas chromatographs is therefore only about 5 vol ppm. Figure 136 shows the determination of 60 vol ppm of acetone and methanol in ethylene, and Fig. 137 illustrates the same analysis

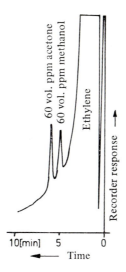

Fig. 136   Analysis of 60 vol ppm of methanol and acetone in 99·99 % pure ethylene
on a 4 m ethylene glycol bis(2-cyanoethyl ether) column at 75°C with a thermal
conductivity detector.

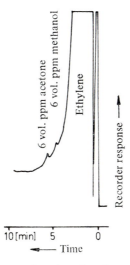

Fig. 137   Analysis of 6 vol ppm of methanol and acetone in 99·99 % pure ethylene
on a 4 m ethylene glycol bis(2-cyanoethyl ether) column at 75°C with a thermal
conductivity detector.

at the level of 6 vol ppm. A thermal conductivity detector was used since the FID has no significant advantage with respect to sensitivity in the case of methanol.[32]

The relatively high limit of detection for methanol is due to both tailing of the ethylene peak and the fact that, at low concentrations, methanol has a broad peak compared with other organic compounds. The reliable determination of traces of methanol is therefore possible with industrial gas chromatographs only when the peak cutting technique is employed.

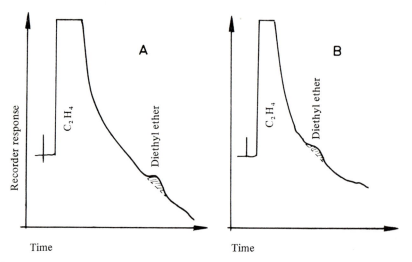

Fig. 138   Determination of traces of diethyl ether in ethylene on dimethylsulpholane (A) and dinonyl phthalate (B) (from J. Sverak and P. L. Reiser, *Mikrochim. Acta*, 163 (1958)).

In the manufacture of ethylene from ethanol, other impurities, *e.g.* diethyl ether, have to be measured. This analysis can also only be carried out in the tail of the ethylene peak.[33] Figure 138 shows the gas chromatographic analysis of traces of diethyl ether in ethylene, and a limit of 30 vol ppm has been reported for this analysis.[34]

PROPYLENE

Most of the propylene used at present is obtained from petroleum refinery gases, and as a raw material for the aliphatic chemicals industry it is no less important than ethylene. It is used for the manufacture of isopropanol, acetone, acrolein and allyl chloride, etc., and as the starting material for oxo-syntheses. It is also of increasing importance for the manufacture of polypropylene, and as a monomer for copolymerizations. These latter uses again require a monomer of high purity. Examination of the boiling points of the gaseous hydrocarbons in Table 9 shows that gas chromatographic separation of compounds present in high-purity propylene will be more difficult than in the case of ethylene, since there are more C_3 than C_2 hydrocarbons to be determined. In addition, the boiling points of the sulphur compounds concerned, *i.e.* H_2S and COS, are closer to that of propylene than ethylene.

The C_2 *hydrocarbons ethane, ethylene and acetylene* can be analysed without any difficulty. Kent has described a rapid method of determining traces of acetylene in propylene for the purpose of plant control with an industrial gas chromatograph.[21] The analysis was carried out with a dimethylsulpholane column and an argon ionization detector, and had a limit of detection of 5 ppm of acetylene in the tail of the propylene peak. However, analysis of this compound and the other C_2 hydrocarbons is simpler if a non-polar stationary phase is used, since all of them can then be eluted before the main component. If a flame ionization detector is used, it is possible to detect less than 1 ppm without any difficulty. Analysis of all the C_2 hydrocarbons in propylene on a perhydrophenanthrene column is illustrated in Fig. 139.[5]

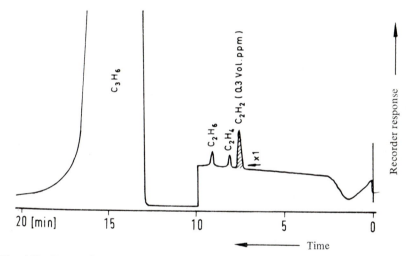

Fig. 139 Determination of traces of acetylene, ethylene and ethane in propylene on a 20 m perhydrophenanthrene column with an FID at 22°C.

Whilst traces of the C_4 compounds present can be detected after the main component, it is not possible to measure the remaining C_3 hydrocarbons, since they are masked by the main peak. *Analysis of the C_3 hydrocarbons propane, cyclopropane, propadiene (allene) and propyne (methylacetylene)* therefore requires special conditions for their separation. The amounts of propyne and propadiene present are of considerable interest in the polymerization of propylene. Analysis of traces of these two compounds is therefore very important. Even 15 years ago it could be carried out down to 5 vol ppm of each compound in propylene[32] (Fig. 140).

As can be seen from Fig. 140, even if an FID is used it is not possible to attain a very low limit of detection for propadiene owing to the overlapping of the peaks. However, propyne can be detected down to 1 ppm. Vigdergauz and Afanas'ev have used a combination of two columns to determine propyne and propadiene in propylene-propane fractions.[35] The stationary phase in column 1 consisted of heptadecane, and in column 2 it was di-isodecyl phthalate. Propyne was eluted between propane and isobutane on column 2. The first column enabled propadiene to be resolved from propylene and isobutane. Chromatographic conditions can be chosen which enable traces

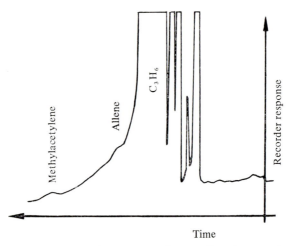

Fig. 140   Analysis of 5 ppm of propyne and propadiene in propylene on a 15 m acetonyl acetone column at 22°C with a thermistor detector (5 ml sample).

of propadiene to be analysed before the main component elutes. Thus, Bua *et al.* have used 30% silver nitrate in ethylene glycol as a stationary phase, and were able (in 1959) to separate the propadiene in front of the propylene.[36] By using a thermal conductivity detector and a 25 ml sample, they measured down to 5 ppm of propadiene, and this could be reduced to 1 ppm by pre-concentration techniques.

The efficiency of this column packing depends on the reversible formation of a complex with the propylene double bond, which increases its retention time considerably, whereas propadiene, with its cumulated double bonds, is not retarded.[37,38] The use of this kind of column packing, in combination with sensitive detectors, *e.g.* the FID, makes it possible to measure propadiene in the ppb region. The amounts of propane and cyclopropane present may be determined at the same time, as shown in Fig. 141.[5]

This analysis is more difficult if it involves crude propylene gas, which can still contain relatively large amounts of C_4 hydrocarbons. Determination of propadiene can then be prevented by *trans*-butylene-2. By combining the

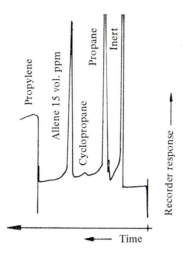

Fig. 141   Determination of traces of allene in propylene on a 4 m silver nitrate–diethylene glycol column at 22°C with an FID. Propylene peak ×500; allene peak ×50.

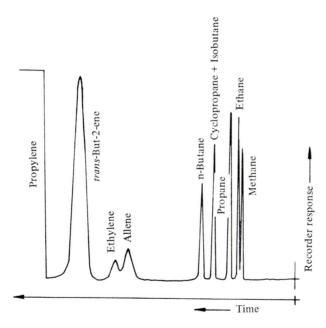

Fig. 142   Determination of traces of propadiene in highly impure propylene with silver nitrate–diethylene glycol + perhydrophenanthrene columns (column lengths 4·2 m/2 m, temperature 22°C).

silver nitrate–diethyl glycol column mentioned above with a perhydrophenanthrene column these two substances may easily be resolved, so that traces of propadiene can be determined even in highly impure propylene (*see* Fig. 142).

The simultaneous determination of cyclopropane is not possible in this instance since it is masked by isobutane, as seen in Fig. 142. Cyclopropane can only be measured by means of differential techniques involving other columns. The stationary phases dimethylsulpholane, hexamethylphosphoramide, methoxyethoxyethyl ether, di-n-decyl phthalate, n-hexadecane, propylene carbonate + dimethylsulpholane, silica gel and squalane provide further possibilities for the analysis of hydrocarbons in propylene.[39]

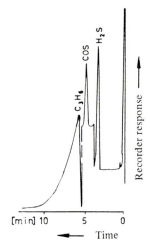

Fig. 143   Analysis of traces of H$_2$S and COS in propylene on a 2 m Porapak Q
column at 22°C.

As a result of the small differences between the boiling points of *hydrogen sulphide* and *carbonyl sulphide* and that of propylene, these two compounds are much more likely to be present in propylene than in ethylene. Nevertheless, their gas chromatographic determination is relatively easy in propylene, since they can be resolved and eluted before the main peak. For example, Watanabe *et al.* measured COS on a combination of Porapak and silica gel columns at 80°C.[40] The analysis of traces of H$_2$S and COS in propylene on Porapak Q is shown in Fig. 143.[32]

If the concentration of COS is below 10 ppm and the column efficiency is low, then it may not be possible to determine COS since its small peak will merge with the propylene peak. A flame photometric detector must be used for this analysis, the operating conditions being the same as for ethylene (*see* Fig. 135).

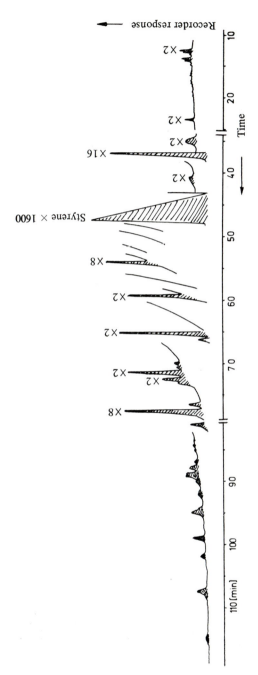

Fig. 144 Gas chromatographic determination of the purity of styrene on a 100 m squalane capillary column at 100°C.

STYRENE

It is generally simpler to carry out trace analyses of liquid monomers since gases frequently involve rather difficult sampling problems. Also, it is easier to make the calibrations. There are very few problems as far as the detector is concerned, since the FID and other ionization detectors can be almost universally used.

Barbul *et al.* have used a packed column and an argon detector to analyse the impurities in styrene.[41] The alkanes and alkenes were eluted before benzene on polyethylene glycol 400. A heavy oil containing no aromatic hydrocarbons was used to separate the aromatics, benzene, toluene, ethylbenzene, *m*- and *p*-xylene, cumene, etc. A third column containing the above phases in a 1:1 ratio resolved *o*-xylene from styrene and cumene. The limits of detection were 10 ppm for benzene, toluene and ethylbenzene, 50 ppm for α-methylstyrene and 100 ppm for benzoic acid. In the case of styrene, it is preferable to employ capillary columns since, when combined with a sensitive detector, almost all impurities present in concentrations exceeding 5 ppm can be determined from a single gas chromatographic separation. Figure 144 illustrates the determination of the purity of a commercial styrene, in which 38 impurities can be distinguished with a 100 m squalane capillary column. For comparison, Fig. 145 shows the chromatogram of the same material on a packed column having squalane as the stationary phase. At best, only 10 impurities can be detected in this case.[32]

As can be seen from the chromatogram on the squalane column, the number of trace components is significantly higher than in ethylene or propylene. The impurities can be identified in these monomers by means of retention data,

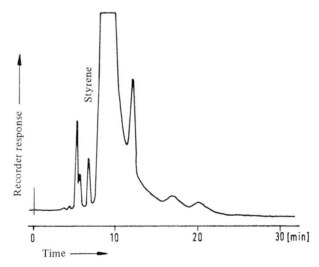

Fig. 145   Determination of the purity of styrene with a 6 m squalane column at 80°C.

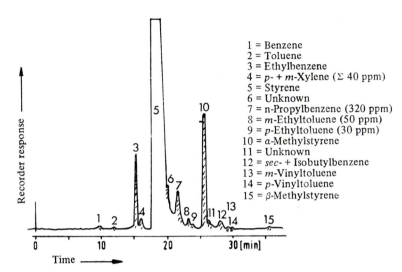

1 = Benzene
2 = Toluene
3 = Ethylbenzene
4 = p- + m-Xylene (Σ 40 ppm)
5 = Styrene
6 = Unknown
7 = n-Propylbenzene (320 ppm)
8 = m-Ethyltoluene (50 ppm)
9 = p-Ethyltoluene (30 ppm)
10 = a-Methylstyrene
11 = Unknown
12 = sec- + Isobutylbenzene
13 = m-Vinyltoluene
14 = p-Vinyltoluene
15 = β-Methylstyrene

Fig. 146 Determination of purity of styrene on a 60 m Apiezon L capillary column at 100°C (from O. L. Hollis, *Anal. Chem.*, **33**, 352 (1961)).

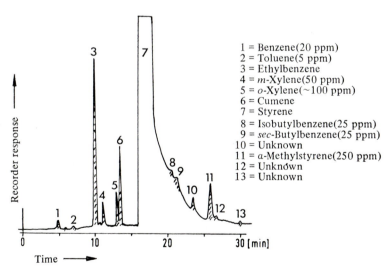

1 = Benzene(20 ppm)
2 = Toluene(5 ppm)
3 = Ethylbenzene
4 = m-Xylene(50 ppm)
5 = o-Xylene(~100 ppm)
6 = Cumene
7 = Styrene
8 = Isobutylbenzene(25 ppm)
9 = sec-Butylbenzene(25 ppm)
10 = Unknown
11 = a-Methylstyrene(250 ppm)
12 = Unknown
13 = Unknown

Fig. 147 Determination of purity of styrene on a 38 m polypropylene glycol capillary column at 70°C (from O. L. Hollis, *Anal. Chem.*, **33**, 352 (1961)).

but this is inadequate for styrene and identification can only be achieved with an efficient g.c.–m.s. combination. Hollis has carried out similar studies with a capillary column and an argon detector.[42] Chromatograms on Apiezon L, polypropylene glycol and bis(phenoxyphenyl) ether capillary columns showed that the best resolution is obtained on the last of these stationary phases. Thus, on the Apiezon L column, which separates predominantly according to boiling points, it is not possible to resolve *p*-xylene from *m*-xylene, *o*-xylene, styrene and cumene, and isobutylbenzene from *sec*-butylbenzene. Gas chromatograms of styrene on these three capillary columns are shown in Figs. 146, 147 and 148.

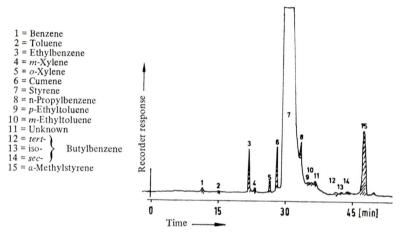

1 = Benzene
2 = Toluene
3 = Ethylbenzene
4 = *m*-Xylene
5 = *o*-Xylene
6 = Cumene
7 = Styrene
8 = n-Propylbenzene
9 = *p*-Ethyltoluene
10 = *m*-Ethyltoluene
11 = Unknown
12 = *tert-*
13 = *iso-* } Butylbenzene
14 = *sec-*
15 = α-Methylstyrene

Fig. 148   Determination of purity of styrene on a 60 m bis(phenoxyphenyl) ether capillary column at 105°C (from O. L. Hollis, *Anal. Chem.*, **33**, 352 (1961)).

ACRYLIC ESTERS

Methyl methacrylate—Haslam and Jeffs have analysed the impurities in a sample of synthetic monomer and one obtained by depolymerization.[43] In the synthetic product, methanol and methyl α-hydroxyisobutyrate were determined, with benzene as an internal standard. The column used for separation contained glycerol and dinonyl phthalate on Celite. Monomer obtained from waste polymethyl methacrylate contained a number of additional impurities which were identified as water, methyl acrylate, methyl propionate and methyl isobutyrate.

Kosek has used two types of column to determine the purity of this monomer.[44] Polyethylene glycol 400 and β,β′-oxydipropionitrile were employed for components having boiling points below 100°C (methyl and ethyl methacrylate, hydrogen cyanide, acetone, methyl formate, methyl acetate, dimethyl ether and methyl, ethyl and propyl alcohols) and Apiezon L to determine the components boiling above 100°C.

Acrylonitrile—Eustache *et al.* have employed mass spectrometry in a determination of the impurities in acrylonitrile.[45] The lower limit of detection was reported to be 1–10 ppm.

Methyl acrylate—The impurities can be separated from the monomer on polyethylene glycol 1000, and the following compounds have been detected: dimethyl ether, formaldehyde, acetaldehyde, methyl formate, methyl propionate, methanol and methyl β-methoxypropionate. The influence of these impurities on the rate of polymerization and molecular weight of the polymer was also studied.[46]

Diethyl ether, ethanol, ethyl acetate, acrylonitrile, ethyl propionate, 2-ethoxyethyl propionate and ethyl acrylate have been determined on a triethylene glycol benzoate capillary column with 2,2,4-trimethylpentane as an internal standard.[47] By silanizing hydroquinones with HMDS they may be determined in acrylic ester monomers.[48] With silicone rubber as the stationary phase and an FID, down to 4 ppm hydroquinone and 9 ppm methoxyhydroquinone can be detected by this technique.

DIVINYLBENZENE

Divinylbenzene is employed as a cross-linking agent in polymers, and is manufactured from diethylbenzene by catalytic dehydrogenation. As a consequence of its tendency to polymerize, it is obtained from the manufacturing process at a maximum concentration of only 60% and is sold in this form. The remaining 40% consists of unconverted diethylbenzene, ethylvinylbenzene and traces of other impurities. The vinyl derivatives and parent compound constitute a mixture of isomers. The trace impurities have been reported to be: 0·1–0·4% naphthalene, 0·01% aldehydes, 0·03% ketones, 0·02% peroxides, 0·02% acetylene compounds and 0·1% *p-t*-butyl pyrocatechol.[42] The stationary phases employed to resolve the isomers have been ethylene glycol bis(β-cyanoethyl ether), polyethylene glycol 3000, tricresyl phosphate, Apiezon L and high vacuum grease R, the last three phases being the most suitable. In other gas chromatographic investigations of the purity of divinylbenzene by Wolf *et al.* and by Wiley and Dyer, an average of 10–20 components was detected.[50,51] Wiley and other co-workers have prepared 99% pure divinylbenzene by preparative gas chromatography.[52] While these reports have been concerned with the separation of the isomers, particularly *m*- and *p*-divinylbenzene, Hannah *et al.* using an FID have analysed the other by-products of divinylbenzene which are present at considerably lower concentrations.[53] Their chromatograms showed 33 components including benzene, toluene, ethylbenzene, cumene, ethyltoluene, styrene, α-methylstyrene, vinyltoluene, 1,2-dihydronaphthalene and naphthalene. The identities of these trace components were confirmed with standard mixtures made up in the same concentration range. The stationary phases used were Carbowax 600 and di-2-ethylhexyl sebacate.

VINYL ACETATE

Vinyl acetate is one of the most important raw materials for the production of aqueous dispersions of polymers. It is manufactured from acetylene and acetic acid by thermal decomposition of ethylidene acetate, or more recently by oxidation of ethylene in the presence of acetic acid. Different impurities have to be considered depending on the manufacturing process employed. Usami has determined acetaldehyde, methyl acetate, acetone, methanol, benzene and crotonaldehyde in the range 60–200 ppm using silicone oil DC 510 and a mixture of stearic acid and glycerol as the stationary phases.[54]

TABLE 10 Retention data for the probable impurities in vinyl chloride on a 10 m tricresyl phosphate column at 35°C (from G. M. Sassu *et al.*, *J. Chromatog.*, **34**, 394 (1968))

Substance	Relative retention time 1,3-*butadiene* = 1·00 at 35°C
Ethane, ethylene	0·05
Acetylene	0·15
Propane	0·17
Propylene	0·22
Isobutane	0·34
Propadiene	0·48
n-Butane	0·53
Butene-1, isobutene	0·64
Methylacetylene	0·68
Butene-2 (*trans*)	0·77
Methyl chloride	0·96
1,3-Butadiene	1·00
Vinyl chloride	—
Vinylacetylene	2·05
Ethyl chloride	2·48
2-Chloropropene-1	2·91
Vinyl bromide	3·28
2-Chloropropane	4·17
1-Chloropropene-1 (*cis*)	4·37
1,1-Dichloroethylene	4·78
1-Chloropropene-1 (*trans*)	5·52
1-Chloropropane	6·88
Diacetylene	7·66
3-Chloropropene-1	8·14
1,2-Dichloroethylene (*trans*)	9·60
Dichloromethane	11·02
2-Chlorobutane	12·05
2-Chloro-1,3-butadiene	12·85
1,1-Dichloroethane	14·22

VINYL CHLORIDE

Vinyl chloride is manufactured either by removal of hydrogen chloride from 1,2-dichloroethane, or by addition of hydrogen chloride to acetylene. The acetylene employed is derived from thermal cracking processes, so that there is the possibility of several hydrocarbon impurities being present. Vinyl chloride produced from acetylene and hydrogen chloride has been analysed by Lazaris by means of a column containing three different stationary phases—2 m polyethylene glycol adipate, 1·5 m dioctyl phthalate and 1·5 m tricresyl phosphate on firebrick.[55] The following compounds were detected: ethylene chloride, 2-chloroprene-1, acetaldehyde, *cis*-1-chloro-1-propene, vinylidene chloride, n-propyl chloride, *trans*-1,2-dichloroethylene, 1,1-dichloroethylene and methanol. Mikkelsen *et al.* have used a combination of two columns and a cold trap to separate the low-boiling from the higher-boiling impurities in the determination of the purity of vinyl chloride.[56] Sassu *et al.* recommend tricresyl phosphate on Chromosorb P for this analysis.[57] With a 10 m column of this material and an FID, all the impurities eluting before and after the main peak can be measured over the concentration range from 0·2 to several ppm. The impurities in vinyl chloride manufactured either from acetylene or from 1,2-dichloroethane can be analysed by this method. Table 10 lists the probable impurities and their relative retention

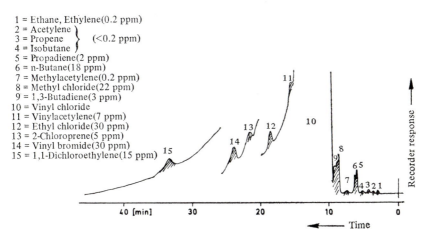

1 = Ethane, Ethylene(0.2 ppm)
2 = Acetylene
3 = Propene   } (<0.2 ppm)
4 = Isobutane
5 = Propadiene(2 ppm)
6 = n-Butane(18 ppm)
7 = Methylacetylene(0.2 ppm)
8 = Methyl chloride(22 ppm)
9 = 1,3-Butadiene(3 ppm)
10 = Vinyl chloride
11 = Vinylacetylene(7 ppm)
12 = Ethyl chloride(30 ppm)
13 = 2-Chloroprene(5 ppm)
14 = Vinyl bromide(30 ppm)
15 = 1,1-Dichloroethylene(15 ppm)

Recorder response

40 [min]   30   20   10   0

← Time

Fig. 149   Gas chromatographic analysis of a commercial grade of vinyl chloride on a 10 m tricresyl phosphate column at 35°C (from G. M. Sassu *et al.*, *J. Chromatog.*, **34**, 394 (1968)).

times on a 10 m tricresyl phosphate column at 35°C. The gas chromatographic analysis of a commercial grade of vinyl chloride on this column at 35°C is illustrated in Fig. 149. In this analysis the hydrogen chloride formed in the flame can cause corrosion of the FID, and Sassu *et al.* therefore recommend that the detector should be constructed of Teflon.

CHLOROPRENE

Chloroprene is produced industrially by the addition of hydrogen chloride to vinylacetylene. Brodsky *et al.* have determined the main impurities, *i.e.* ethylene, vinylacetylene, acetaldehyde, divinylacetylene and vinyl chloride,[58] the analysis being carried out with butane-2,3-diol coated SiO_2 columns of two different lengths (74 cm and 115 cm).

1,3-BUTADIENE

The impurities in 1,3-butadiene depend on the manufacturing process used. This may be either removal of water from 1,3-butylene glycol or ethanol, or dehydrogenation of saturated C_4 hydrocarbons. Because of the number of impurities which may be present, there are considerable problems of separation if detection of each of them is required. For example, 15 columns have been proposed for the analysis of the purity of butadiene.[59]

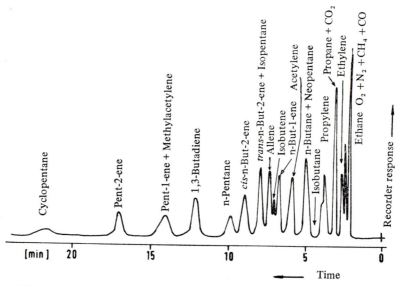

Fig. 150   Separation of C_2–C_5 hydrocarbons on a 15 m acetonyl acetone column at 21°C.

Higgins and Baldwin have presented a comprehensive survey of the separation of C_4 hydrocarbons.[60] Additional columns which may be used for the analysis of butadiene are: hexamethylphosphoramide on firebrick, bis-2(2-methoxyethoxy)ethyl ether on Chromosorb, *N,N'*-dimethylformamide on Al_2O_3 and propylene carbonate on Al_2O_3.[61–64] 1,3-Butadiene can be satisfactorily separated from all other C_2–C_5 hydrocarbons with acetonylacetone on Sterchamol, as shown in Fig. 150.[65] If a flame ionization detector is used this column must be operated at 0°C.

An elegant method of separating this monomer from its impurities consists of using maleic anhydride as stationary phase. This results in a considerable increase in retention time of the 1,3-butadiene compared with the other constituents.[66-68]

PHTHALIC ANHYDRIDE

Phthalic anhydride is one of the raw materials for the manufacture of polyesters and alkyd resins. In its manufacture by the oxidation of naphthalene or o-xylene, a number of impurities are formed which can subsequently cause discolouration or odours. Trachman and Zucker have described a direct method of determining these impurities by dissolving the anhydride in o-dichlorobenzene, and injecting samples of this solution into the gas chromatograph.[69] A silicone SF 96 column was used at 220°C to determine

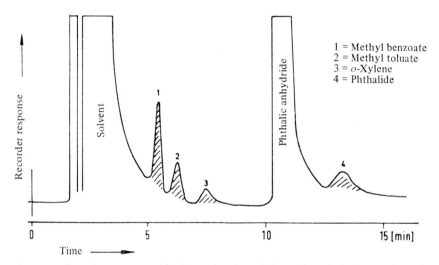

Fig. 151   Gas chromatographic determination of the purity of phthalic anhydride using silicone grease at 185°C as stationary phase (from C. E. Moore and S. Meinstein, *Anal. Chem.*, **34**, 1503 (1962)).

o-dichlorobenzene, benzoic acid, naphthalene, maleic anhydride and 1,4-naphthaquinone. Traces of benzoic acid and o-toluic acid may be analysed in a benzene–ether solution after they have been esterified with diazomethane.[70] The chromatogram in Fig. 151 illustrates this analysis with a β-ionization detector and silicone grease at 185°C as stationary phase. Efficient separation of these methyl esters can also be achieved with silicone rubber and Carbowax 20M on acid-washed Chromosorb W.[71] In this instance, esterification was carried out with methanol and sulphuric acid, since it was suspected that diazomethane can form pyrazoline derivatives with the methyl ester of maleic acid. The column temperature was 175°C at the beginning of the analysis, and was subsequently programmed to 200°C. The following compounds were resolved: o-xylene, dimethyl ester of maleic acid, dimethyl ester of citraconic acid, methyl benzoate, methyl o-toluate, naphthalene, methyl ester of phenoxyacetic acid and dimethyl phthalate.

FORMALDEHYDE

Formaldehyde is a monomer which is repeatedly a source of problems in gas chromatography. This applies to both the analysis of traces of formaldehyde in gases, liquids and solids, and of the impurities in formaldehyde itself.

Gaseous and liquid formaldehyde can only be stored if certain precautions are taken, since it readily forms polyoxymethylene. The monomer is normally available in the form of its aqueous solution, 'formalin', which is stabilized by the addition of alcohols, usually methanol, for storage and transportation. The equilibrium for the reaction:

$$CH_2O + H_2O \rightleftharpoons HOCH_2OH$$

lies almost completely on the side of the monohydrate, methylene glycol, from which polyoxymethylene dihydrates having the formula $HO(CH_2O)_nH$ are formed by polycondensation reactions. Those hydrates for which $n = 1–3$ are soluble in water, but higher oligomers are no longer soluble, and form the familiar white deposit of 'paraformaldehyde' in formalin solutions.[72]

Other impurities which have to be considered are methyl formate, and formic acid, which can be formed both by the Cannizzaro reaction

$$2CH_2O + H_2O \rightarrow CH_3OH + HCOOH$$

and by atmospheric oxidation of formaldehyde. Moreover, dimethylene glycol $(CH_3OCH_2OCH_3)$ and other oligomers, the acetals: $CH_3O(CH_2O)_{2-3}CH_3$ and the corresponding hemiacetals: $CH_3O(CH_2O)_{1-3}H$ may also be present. Whereas formaldehyde can easily be estimated in aqueous solution by titration, determination of the other impurities mentioned above, especially when present in trace amounts, is very rarely possible by wet chemical methods. Thus, for example, the titrimetric determination of low concentrations of formic acid in aqueous formalin solution give incorrect results, since the solution itself has an acidic reaction:

$$HO(CH_2O)_{1-3}H \rightleftharpoons HO(CH_2O^-)_{1-3}H^+$$

and consequently gives a misleading value for formic acid.

Ever since gas chromatography was developed attempts have been made to use it to analyse aqueous solutions of formaldehyde. However, the results obtained are often so inconsistent that it is extremely difficult to obtain a clear idea of the nature and concentrations of the impurities. Those who have to attempt this analysis will rapidly appreciate the problems involved. Thus, separation of formaldehyde from water and methanol is complicated on most gas chromatographic columns. The direct gas chromatographic

analysis of traces of formic acid is also rather difficult, since a flame ionization detector cannot be used. In addition, unavoidable memory effects
sometimes occur with this polar substance, and this also applies to the
determination of water. Nevertheless, the most inconsistent results occur in
the determination of formaldehyde itself, since this substance is not only
highly polar, but tends to polymerize most readily. Consequently, in many
investigations, the formaldehyde peak has a very broad shape and is eluted
at unexpected retention times. On Ethofat 60/25 (polyoxymethylene mono-
stearate), formaldehyde is eluted after water.[73] Columnpak T is recommended
as the most suitable column packing for separation of the components
present in formalin solution. The column temperature should be 115°C,
and the injection block at 225°C. Under these conditions both for aqueous

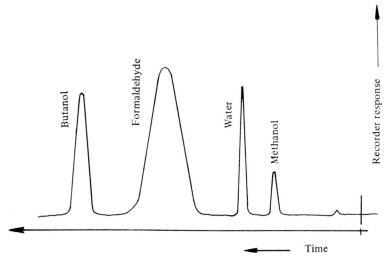

Fig. 152   Determination of formaldehyde on Columnpak T at 115°C (from K. J.
Bombaugh and W. C. Bull, *Anal. Chem.*, **34**, 1237 (1962)).

solutions and gaseous formaldehyde, the monomer is eluted relatively late.
The very broad peak for formaldehyde and its boiling point of −19°C are not
consistent with its retention time compared with the other components, as
Fig. 152 shows.

Similar results have been obtained by Stevens and by Gavrilina and
Natalina.[74,75] The anomalous elution of formaldehyde can probably be
attributed to the equilibrium between polar formaldehyde and the stationary
phase being established very slowly. There is also the obvious possibility
that it is not formaldehyde, but methylene glycol, which produces this broad
peak, due to the following equilibrium being established:

$$CH_2O + H_2O \rightleftharpoons HOCH_2OH$$

This suggestion is based on the observation that the formaldehyde peak is not broad if it is eluted before water, as is the case with sucrose octa-acetate on Columnpak T, Citroflex A8 (*O*-acetyltriethylhexyl citrate), pentaerythritol tetra-acetate on Phase Sep universal A, or Phase Pak Q, polyethylene glycol adipate on Celite, and Carbowax 20M on Chromosorb.[76–80] Successful gas chromatographic analysis may also be achieved with Porapak N[81] (*see* Fig. 153).

The column temperature for the analysis shown in Fig. 153 was 120°C, and the vaporizer and thermal conductivity detector were at 240°C. The quantitative results of this gas chromatographic determination were in good agreement with those from a titration method.

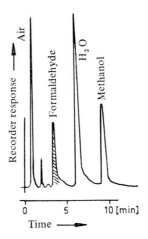

Fig. 153   Determination of formaldehyde and the impurities in its aqueous solution on Porapak N (from F. Onuška and J. Janák, *J. Chromatog.*, **40**, 209 (1969)).

Gruber and Plainer have undertaken detailed studies of the anomalous gas chromatographic behaviour of formaldehyde, and the nature of its impurities.[82] The asymmetric peak which appears in the analysis of methanolic formaldehyde solutions was believed to be formaldehyde–methylhemiacetal, which should have attained equilibrium of formation and dissociation. They concluded that it is not a single chemical species which is eluted. The other compounds present were identified by comparing their retention times with reference compounds on Carbowax 1500 and on Porapak, both a thermal conductivity and flame ionization detector being used. Valuable information about the identity of these other compounds was also obtained by analysis of the individual components isolated at the exit of the gas chromatograph. Methanol, methylal, di- and tri-oxymethylene dimethyl ethers and dimethyl ether were identified in this way, but methyl formate and formic acid could not be detected.

In accord with Onuška, Gruber and Plainer also recommended Porapak N as the most suitable column packing for the gas chromatographic analysis of formaldehyde, and pointed out that the column should be operated at a temperature exceeding 100°C.

Special mention should also be made here of the head space method, which can sometimes give very useful information in the analysis of aqueous formaldehyde solutions (Gruber).

References
1. STEVENS, M. P., *Characterization and analysis of polymers by gas chromatography*, Marcel Dekker, New York and London (1969).
2. WINNACKER, K. and KÜCHLER, L., *Chem. Technol.*, Volume 3, p. 716, Carl Hauser Verlag, Munich (1950).
3. RAY, N. H., *Analyst*, **80**, 853 (1955).
4. JANÁK, J., *Coll. Czech. Chem. Comm.*, **18**, 798 (1953).
5. HACHENBERG, H. and GUTBERLET, J., *Brennstoff-Chem.*, **49**, 279 (1968).
6. PIETSCH, H., *Erdöl Kohle*, **11**, 157 (1958).
7. LITYAEVA, Z. A., MARKOSOV, P. I. and ZAICHENKO, V. N., *Tr. Neftegaz, Nachn. Issled. Inst. Krasnodarsk Filial*, 110 (1962).
8. BRENNER, N. and ETTRE, L. S., *Anal. Chem.*, **11**, 1815 (1959).
9. NODOP, G., *Z. Anal. Chem.*, **164**, 120 (1958).
10. ALEKSEEVA, A. V. and GOL'BERT, K. A., *Zavodsk. Lab.*, **27**, 972 (1961).
11. ZLATKIS, A. and KAUFMAN, H. R., *J. Gas Chromatog.*, **4**, 240 (1966).
12. VAN LANGERMEERSCH, A., *Proc. 5th World Petrol. Congr.*, Section V, 65–74.
13. GOL'BERT, K. A. and ALEKSEEVA, A. V., *Zavodsk. Lab.*, **24**, 688 (1958).
14. HARA, N., 15th Annual meeting of the Chemical Society of Japan.
15. ALEKSEEVA, A. V., GOL'BERT, K. A. and FOMINA, A. I., *Neftekhimiya*, **5**, 449 (1965).
16. ALEKSEEVA, A. V., BERMAN, S. S., GOL'BERT, K. A., PATSKEVITCH, A. A., MOSHINSKAYA, M. B. and FOMINA, A. I., *Gaz. Khromatogr. Akad. Nauk. SSSR. Tr. Storoi Vses*, Moscow 99 (1962).
17. SCHWENK, U., HACHENBERG, H. and FÖRDERREUTHER, M., *Brennstoff-Chem.*, **42**, 194 (1961).
18. DERST, P., VAN VESSEM, I. L. and WICHLER, E., *Dechema Monographien* **62**, 185 (1968).
19. AL'PEROVICH, V. Ya., *Zavodsk. Lab.*, **30**, 1317 (1964).
20. PAYLOR, R. A. L. and FEINLAND, R., *Anal. Chem.*, **33**, 808 (1961).
21. KENT, T. B., *Chem. Ind.* (London), 1260 (1960).
22. RENNHAK, S., DÖRING, C. E., SCHMID, G., SCHNELLER, D., STÜRTZ, H. and WERNER, E., *Chem. Tech.* (Berlin), **17**, 688 (1965).
23. BRADFORD, B. W., HARVEY, D. and CHALKLEY, D. E., *J. Inst. Petrol.*, **41**, 80 (1955).
24. HACHENBERG, H. and GUTBERLET, J., *Brennstoff-Chem.*, **49**, 279 (1968).
25. VIGDERGAUZ, M. S. and ANDREEV, L. V., *Zavodsk. Lab.*, **5**, 550 (1965).
26. KAISER, R., *Chromatographia*, **2**, 453 (1969).
27. KAISER, R., *Chromatographia*, **3**, 38 (1970).
28. ZLATKIS, A., KAUFMAN, H. R. and DURBIN, D. E., *J. Chromatog. Sci.*, **8**, 416 (1970).
29. TABUTEAU, J., *Ind. Chem. Belge*, Suppl. **1**, 138 (1959).

30. ZLATKIS, A., CHAO, G. S. and KAUFMAN, H. R., *Anal. Chem.*, **36,** 2354 (1964).
31. HOLLIS, O. L. and HAYES, W. V., *6th International symposium on GC, Rome*, (Ed. A. B. Littlewood), Paper 8, Inst. of Petrol. (1966).
32. HACHENBERG, H., unpublished work.
33. RAY, N. H., *J. Appl. Chem.*, **4,** 21 (1954).
34. SVERAK, J. and REISER, P. L., *Mikrochim. Acta*, 159 (1958).
35. VIGDERGAUZ, M. S. and AFANAS'EV, M. I., *Zh. Analit. Khim.*, **19,** 1122 (1964).
36. BUA, E., MANARESI, P. and MOTTA, L., *Anal. Chem.*, **31,** 1910 (1959).
37. VAN DE CRAATS, F., *Anal. Chim. Acta*, **14,** 136 (1956).
38. BEDNAS, M. E. and RUSSELL, D. S., *Can. J. Chem.*, **36,** 1272 (1958).
39. *Manual on hydrocarbon analysis*, 2nd Edition, p. 543. Published by the American Society for Testing and Materials, 1916 Race St., Philadelphia, Pa 19103 (1968).
40. WATANABE, Y., ISOMURA, K. and TOMARI, Y., *Japan Analyst*, **16,** 942 (1967).
41. BARBUL, M., POP, A. and BESCHEA, C., *Rev. Chim.* (Bucharest), **15,** 280 (1964).
42. HOLLIS, O. L., *Anal. Chem.*, **33,** 352 (1961).
43. HASLAM, J. and JEFFS, A. R., *J. Appl. Chem.*, **7,** 24 (1957).
44. KOSEK, B., *Chem. Prumysl*, **15,** 160 (1965).
45. EUSTACHE, H., GUILLEMIN, C. I. and AURICOURT, F., *Bull. Soc. Chim. France* (5), 1386 (1965).
46. VYAKHIREV, D. A., ZABOTIN, K. P., ZUEVA, E. M., TROITSKII, B. B. and FOMICHEVA, L. V., *Tr. po Khim. i Khim. Tekhnol.*, (2), 279 (1964).
47. LYUTOVA, T. M. and LAZARIS, A. Y., *Zh. Analit. Khim.*, **21,** 1146 (1966).
48. EHRLICH, M. H. Jr, Pittsburgh Conf. on Anal. Chem. and Appl. Spectry, Cleveland/Ohio, 1st–6th March, 1970, Abstr. Papers No. 278.
49. BLASIUS, E. and BEUSHAUSEN, J., *Z. Anal. Chem.*, **197,** 228 (1963).
50. WOLF, F. *et al., Abhandl. Deut. Akad. Wiss. Berlin, Kl. Chem. Geol. Biol.*, No. 1, 513 (1962).
51. WILEY, R. H. and DYER, R. M., *J. Polymer Sci.*, Part A, No. 2, 3153 (1964).
52. WILEY, R. H., DEVENUTO, G. and VENKATACHALAM, T. K., *J. Gas Chromatog.*, **5,** 590 (1967).
53. HANNAH, R. E., COOK, M. L. and BLANCHETTE, J. A., *Anal. Chem.*, **39,** 358 (1967).
54. USAMI, S., *Bunseki Kagaku*, **10,** 141 (1961).
55. LAZARIS, A. Ya., *Neftekhimiya*, **5,** 166 (1965).
56. MIKKELSEN, L., SPENCER, S. and SZYMANSKY, H. A., *Lectures on GC* 1962, Plenum Press, New York (1963).
57. SASSU, G. M., ZILIO-GRANDI, F. and CONTE, A., *J. Chromatog.*, **34,** 394 (1968).
58. BRODSKY, J., MAČKA, M. and MIKL, O., *Chem. Prumysl*, **10,** 460 (1960).
59. ASTM Designation: D 2593-67 T.
60. HIGGINS, C. E. and BALDWIN, W. H., *Anal. Chem.*, **36,** 473 (1964).
61. McEVEN, D. J., *Chem. Can.*, **11,** 35 (1959).
62. LOYD, R. J., AYERS, B. O. and KARASEK, F. W., *Anal. Chem.*, **32,** 698 (1960).
63. HARA, N., SHIMADA, H., ISHIKAWA, A. and DOHI, K., *Bull. Japan. Petrol. Inst.*, **2,** 33 (1960).
64. McKENNA, T. A. and IDLEMAN, J. A., *Anal. Chem.*, **32,** 1299 (1960).
65. HORN, O., SCHWENK, U. and HACHENBERG, H., *Brennstoff-Chem.*, **39,** 336 (1958).

66. DEVYATYKH, G. G., ZORIN, A. D. and EZHELEVA, A. E., *Nauchn. Dokl. Vysshei Shkoly, Khim. i. Khim. Tekhnol.*, 724 (1958).
67. JANÁK, J. and NOVAK, J., *Chem. Listy*, **51**, 1832 (1957).
68. JANÁK, J. and NOVAK, J., *Coll. Czech. Chem. Comm.*, **24**, 384 (1959).
69. TRACHMAN, H. and ZUCKER, F., *Anal. Chem.*, **36**, 269 (1964).
70. MOORE, C. E. and MEINSTEIN, S., *Anal. Chem.*, **34**, 1503 (1962).
71. CUCARELLA, M. C. M. and CRESPO, F., *J. Gas Chromatog.*, **6**, 39 (1968).
72. WALKER, J. F., *Formaldehyde Monograph Series*, No. 159, 3rd Edition, p. 53, Reinhold Publishing Corp.
73. BOMBAUGH, K. J. and BULL, W. C., *Anal. Chem.*, **34**, 1237 (1962).
74. STEVENS, R., *Anal. Chem.*, **33**, 1126 (1961).
75. GAVRILINA, L. Y. and NATALINA, N. N., *Anal. Abstr.*, **13**, 2994 (1966).
76. MANN, R. S. and HAHN, K. W., *Anal. Chem.*, **39**, 1314 (1967).
77. KELKER, H., *Z. Anal. Chem.*, **176**, 3 (1960).
78. JONES, K., *J. Gas Chromatog.*, **5**, 432 (1967).
79. KOLOUSKOVA, V. and FELTL, L., *Z. Anal. Chem.*, **202**, 262 (1964).
80. SCHEPARTZ, A. I. and McDOWELL, P. E., *Anal. Chem.*, **32**, 723 (1960).
81. ONUSKA, F., JANÁK, J., DURAS, S. and KRČMÁROVÁ, *J. Chromatog.*, **40**, 209 (1969).
82. GRUBER, H. L. and PLAINER, H., *Chromatographia*, **3**, 490 (1970).

2.52. Determination of Residual Monomers and Other Volatile Impurities in Polymers

The quality of a synthetic polymer and its utilization depend to a great extent on the amount of low molecular weight material which it contains. Thus, the residual concentration of monomers, oligomers, solvents and polymerization catalysts affects the resistance to ageing, or they act as undesirable plasticizers. Other compounds such as anti-oxidants, ultraviolet absorbers or plasticizers are added in low concentrations in order to improve the properties of the polymer. Some of these substances impart an unpleasant odour which is undesirable when the polymer is used in decorative paints and floor coverings, or in the packaging of foodstuffs, where the taste may be affected. It is therefore necessary to know the nature, and concentrations, of these additives. Although solid materials are involved, gas chromatographic analysis can be of considerable value.

The main problem with polymers is simply to separate the additives, if they are volatile, to make them available for gas chromatography. There are various ways of approaching this:

(a) Dissolution of the polymer in a suitable solvent and injection of samples of this solution into the gas chromatograph. The solvent chosen should have a retention time which differs as much as possible from those of the compounds being determined, so that the solvent peak does not overlap those of the trace components. The solvent must also be of very high purity. In addition, it must be readily available, relatively inexpensive and easily purified. For these reasons the choice is rather limited. This technique of sampling the polymer solution has the disadvantage that the non-volatile polymer contaminates the injection unit. Moreover, the temperature of the injection block must not be too high, since pyrolysis can produce decomposition products which were not originally present in the sample. This is particularly true of metal injection systems, and gas chromatographs used for this type of analysis should have interchangeable glass, or preferably silica, inserts in the injection block.

(b) Precipitation of the polymer from a solution. This separates from the polymer the low molecular weight constituents to be analysed. Subsequent sampling is therefore straightforward, but the peaks due to the solvent and precipitant frequently interfere with the analysis of the trace components. Moreover, there is the possibility that adsorption or absorption by the precipitated polymers may affect the quantitative determination of the low molecular weight components. In addition, the process of dissolving and reprecipitating the polymer results in dilution, and consequently a significant reduction in sensitivity.

(c) The polymer can be heated at normal or reduced pressure and the volatile constituents collected in a cold trap, or they can be fed directly into the gas chromatograph. This method has to be used for polymers which are only slightly soluble, or completely insoluble. Its disadvantages are that it is relatively slow for routine applications, and is insufficiently reproducible for quantitative analysis.

(d) Head space analysis should be mentioned since it is the most sensitive method for determination of volatile low molecular weight components in polymers, and will be increasingly used in the future (see 2.15).

DETERMINATION OF RESIDUAL MONOMERS AND OTHER IMPURITIES IN POLYSTYRENE

Pfab and Noffz have used two techniques to determine styrene monomer and other aromatics.[1] In the first of these, the polymer was dissolved in o-dichlorobenzene, and a low boiling point fraction, containing all the styrene and other aromatics, distilled off with a spinning-band column. Figure 154 shows the gas chromatographic analysis of a typical distillate.

The low molecular weight constituents can be almost completely removed from polystyrene with petroleum ether, and addition of known amounts of the monomer have demonstrated that absorption of residues does not occur in this method of analysis. The losses during distillation are negligible since they amount to only a few hundredths of one percent, based on polystyrene.

The second technique depends on the precipitation of polystyrene from methylene chloride solution by methanol. The supernatant solution, to which an internal standard (1-phenylbutane) has been added, contains the styrene and other low molecular weight constituents. As previously mentioned, the sensitivity of the method is reduced by dilution with the solvent and precipitant, and consequently a flame ionization detector should be used. The analysis is illustrated in Fig. 155 (see also 2.15, Fig. 89).

This precipitation technique, which was developed by Eisenbrand and Eich, has the disadvantage that the components being analysed usually have to

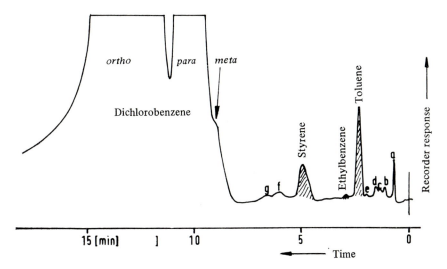

Fig. 154 Determination of styrene and other aromatics in polystyrene (4 m polyglycol column, thermal conductivity detector) (peaks *a* to *g* are impurities in the
solvent) (from W. Pfab and D. Noffz, *Z. Anal. Chem.*, **195**, 37 (1963)).

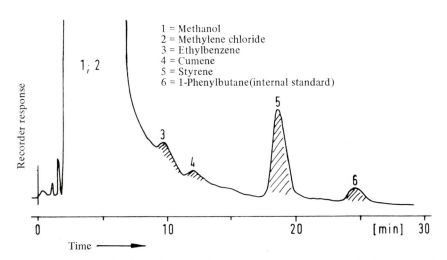

Fig. 155   Determination of styrene in polystyrene by the precipitation method
(2 m polyglycol column, flame ionization detector) (from W. Pfab and D. Noffz,
Z. Anal. Chem., **195**, 40 (1963)).

be measured in the tails of the peaks owing to the solvent and precipitant.[2] Since small quantities of the monomeric form of the precipitated polymer are always absorbed, care should be taken when using an internal standard to ensure that it is added before the polymer is precipitated. Errors of up to 30% can arise if the internal standard is not added until after precipitation.

The results from both of the techniques just described are generally in good agreement, although occasionally the result given by the first method may be somewhat lower, owing to distillation losses.

Residual monomers in copolymers of styrene with α-methylstyrene, acrylonitrile and methyl methacrylate have been determined by Kleshcheva *et al.* by dissolving in chloroform and precipitating with methanol.[3] The gas chromatographic analysis of the residual styrene concentration was carried out on polyethyleneglycol adipate with an argon detector. Isopropylbenzene was employed as the internal standard.

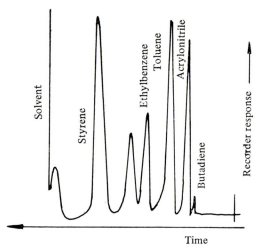

Fig. 156 Analysis of residual monomers in a styrene–acrylonitrile–butadiene terpolymer (column: 1 m Tween on Chromosorb W + 3 m diethyleneglycol succinate on Chromosorb W at 120°C) (from P. Shapras and G. C. Claver, *Anal. Chem.*, **36**, 2282 (1964)).

Shapras and Claver have used a method involving direct injection of solutions of polymers in N,N'-dimethylformamide, with toluene as internal standard, since this solvent can be obtained gas chromatographically pure by distillation.[4] The constituents being analysed, *i.e.* 1,3-butadiene, acrylonitrile, ethylbenzene and styrene were well resolved from each other, and eluted before the solvent (Fig. 156). Additives, pigments and stabilizers do not interfere, but it should be pointed out that periodic cleaning of the sampling unit is absolutely necessary. The method may be used to analyse polystyrene, styrene–acrylonitrile and styrene–butadiene copolymers and also styrene–acrylonitrile–butadiene terpolymers with a limit of detection of less than 10 ppm of the respective monomers.

A comprehensive survey of gas chromatographic methods of determining residual monomers in materials based on polystyrene has been presented by Simpson.[5]

Eggertsen and Stross have used a conventional pyrolysis unit to heat the polymers to below their normal pyrolysis temperature, in their method of analysing acrylonitrile and styrene on polypropylene glycol and silicone oil columns.[6] In the case of the silicone column, adsorption of acrylonitrile was reduced by saturating the carrier gas with water vapour. In a similar method, Crompton and Myers have determined styrene, several aromatics, and other volatile components in the head space vapours produced from various grades of polystyrene by heating them for 15 minutes at 200°C.[7]

Polystyrene also contains saturated and unsaturated styrene oligomers formed during polymerization, and which can have undesirable plasticizing effects. The dimeric styrenes can be separated at 220°C on a silicone oil column.[8] The other oligomers may be analysed with an SE 30 silicone rubber on Chromosorb W column, temperature programmed from 100 to 130°C.[9] The samples for these analyses can be prepared by dissolving polystyrene in benzene and then precipitating the high molecular weight fraction with methanol. By repeated dissolution and reprecipitation of the polymer it has been proved that the oligomers present in polystyrene are completely soluble in methanol, and that none have a degree of polymerization exceeding 5. Stein and Mosthaf have separated 15 of these components on a capillary column.[10] They were identified with a g.c.–m.s. system, and it was shown that four of the peaks can be assigned to dimers of styrene, and four others were due to trimers. The analyses were performed on a 50 m high vacuum grease capillary column, temperature programmed from 170 to 260°C. Under similar gas chromatographic conditions, Kurze *et al.* have detected 12 components consisting of dimers and trimers.[11] These were also identified without ambiguity by their mass spectra. Furthermore identification was confirmed by synthesizing the various possible dimers and trimers and measuring their retention indices.

VOLATILE SUBSTANCES IN POLYETHYLENE

Since polyethylene is more or less insoluble in organic solvents, the low molecular weight impurities cannot be determined in solution, or by dissolving and reprecipitating the polymer. Solvent extraction techniques often involve loss of low boiling components before they can be analysed. However, a possible method of concentrating them in polyethylene consists of removal from the polymer by volatilization at high temperatures (125–200°C). For this purpose, Crompton and Myers have developed an apparatus based on the principle of head space analysis, and which is just as practicable as simpler

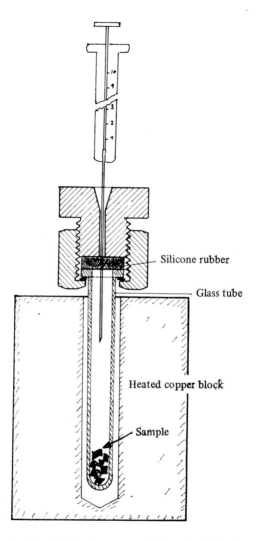

Fig. 157   Apparatus for the head space analysis of solid polymers (from T. R. Crompton and L. W. Myers, *Plastics and Polymers*, 205 (1968)).

units.[7] The apparatus consists of a heated copper block containing a cylindrical glass vessel which holds the sample. The glass tube is sealed with a rubber septum, and the sample can be withdrawn with a normal syringe and injected into the gas chromatograph (*see* Fig. 157).

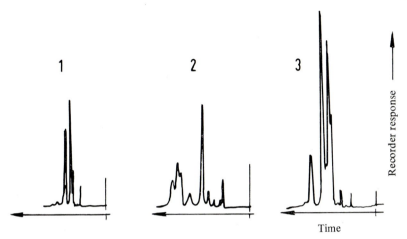

Fig. 158   Head space chromatograms for various polyethylenes (heating period: 15 min at 125°C, column: 60 m copper capillary containing dibutyl phthalate, column temperature: 30°C) (from T. R. Crompton and L. W. Myers, *Plastics and Polymers*, 205 (1968)).

Figure 158 shows gas chromatograms for the head space vapours of polyethylene samples from different manufacturers. The samples were heated for 15 minutes at 125 or 200°C.

Although the peaks were not indentified for these samples, the method nevertheless provides a reproducible and rapid finger-print technique for

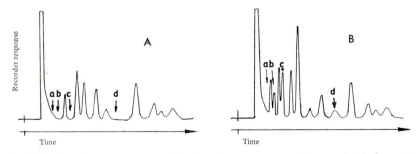

Fig. 159   Gas chromatograms of the head space vapours from polyethylene contaminated with substances producing an odour (*B*) and an odourless sample (*A*). (Heating period: 15 min at 200°C, column: 60 m copper capillary containing dibutyl phthalate, column temperature: 30°C) (from T. R. Crompton and L. W. Myers, *Plastics and Polymers*, 205 (1968)).

comparing and characterizing polyethylenes from different manufacturers. Figure 159 shows a similar comparison of a polyethylene contaminated with compounds producing an odour (*B*) and an odourless sample (*A*). It can be inferred from the two chromatograms that the components *a*, *b*, *c* and *d* are probably responsible for the odour.

FORMALDEHYDE, PHENOL AND OTHER IMPURITIES IN PHENOL–FORMALDEHYDE RESINS

Determination of impurities in phenol–formaldehyde resins is generally limited to formaldehyde and phenol, since these impart the familiar odour to the polymer. Preparation of samples for gas chromatographic analysis involves dissolving the resin in alkali and precipitating with acid. The aqueous phase is then analysed for formaldehyde (*see* 2.51), and its ether extract for phenol.[12] Figure 160 shows the determination of formaldehyde, and Fig. 161 that of phenol.

This method can also be applied to phenolic resins modified with acrolein or furfural, and to phenol–furfural and phenol–acrolein resins.[13]

An alternative method of determining traces of formaldehyde and other carbonyl compounds in aqueous media is the gas chromatographic analysis of their thermally stable 2,4-dinitrophenylhydrazones.[14] After precipitation from 0·2N aqueous HCl, samples of the 2,4-dinitrophenylhydrazones were prepared in chloroform solution. Compared with the direct determination of formaldehyde, this method has the advantage that an FID can be used resulting in a significantly higher sensitivity towards formaldehyde. However,

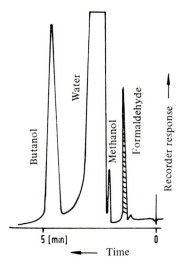

Fig. 160 Determination of formaldehyde on a 4·8 m column containing 10% sucrose octa-acetate on Teflon 6 (column temperature: 130°C) (from M. P. Stevens and D. F. Percival, *Anal. Chem.*, **36**, 1023 (1964)).

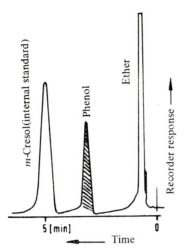

Fig. 161 Determination of phenol on a 3·6 m column containing 10% silicone SF 96 on Fluorapak (column temperature: 130°C) (from M. P. Stevens and D. F. Percival, *Anal. Chem.*, **36**, 1023 (1964)).

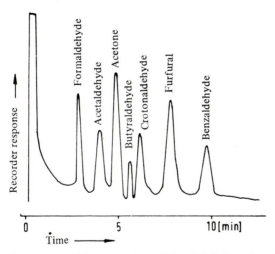

Fig. 162 Gas chromatographic separation of the 2,4-dinitrophenylhydrazones of acetone and some aldehydes (column: 5% SE 30 silicone grease on Chromosorb G, length: 0·7 m, temperature programme: 180–250°C at 10°C min^{-1}, injection temperature: 285°C) (from R. E. Leonhard and J. E. Kiefer, *J. Gas Chromatog.*, **4**, 143 (1966)).

when only trace amounts are present, quantitative precipitation of the formaldehyde–2,4-dinitrophenylhydrazone is very difficult. A gas chromatogram for the determination of acetone and a number of aldehydes is shown in Fig. 162.

A similar technique may be used to analyse phenols in the form of their derivatives, *e.g.* trimethylsilyl ethers or the trifluoroacetate esters.[15,16]

CAPROLACTAM AND OLIGOMERS IN POLYAMIDES

Residual amounts of caprolactam and oligomers greatly affect the mechanical properties of nylon, *e.g.* the viscosity of the molten polymer. Gas chromatographic determination of the concentration of caprolactam in aqueous extracts of nylon may be carried out with an SE 30 silicone rubber stationary phase on Diaport S at 195°C. It is necessary for the internal standard to be soluble in water and bis-[2-(2-methoxyethoxy)ethyl] ether was therefore used.[17] Since the time required to extract the polymer with water is disproportionately long compared with that for the actual gas chromatographic

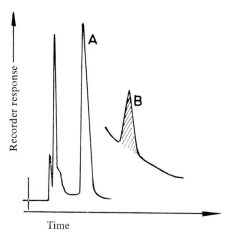

Fig. 163   Determination of residual amounts of caprolactam (peak *B*) in nylon 6 on Carbowax 20 M at 200°C (from F. Zilio-Grandi *et al., Anal. Chem.*, **41**, 1847 (1969)).

analysis, it has been recommended that the polymer should be dissolved in 85% formic acid, and this solution sampled directly.[18] The application of this method to the determination of 0·4% caprolactam (peak *B*) in nylon 6 with an FID and Carbowax 20M as the stationary phase at 200°C is shown in Fig. 163 (peak *A* is the internal standard).

The use of formic acid as solvent has the great advantage that there is little response to it by the FID. Moreover, oligomers do not interfere with the analysis, since they are not eluted under the conditions described. The limit of detection reported for caprolactam is 0·1%. Although the time required for the actual chromatography is only 5–6 minutes, the over-all analysis time depends on the preparation of the sample, *i.e.* on the rate at which the polymer dissolves in formic acid. The oligomers in polyamides (nylon 6, nylon 66 and nylon 12) were determined in the form of their cyclic mono-amines by Mori *et al.*[19] Lithium aluminium hydride was used as the reducing agent, and gas chromatographic analysis was carried out on neo-pentyl glycol succinate and SE 30 silicone rubber columns, both of which contained 3% KOH. Dimers, trimers, tetramers, pentamers, hexamers, caprolactam and 6-aminohexanol were resolved, and lauryl lactam was used as the internal standard.

DETERMINATION OF RESIDUAL MONOMERS IN AQUEOUS DISPERSIONS OF POLYMERS

Knowledge of the amounts of monomers remaining in emulsions of synthetic polymers is of importance with respect to their stability, and because they cause odours in products manufactured from them. The analysis of impurities in aqueous dispersions of polymers is subject to considerable errors, since additionally large amounts of water are present. There is always some uncertainty as to the extent to which the trace components being analysed are absorbed by the polymers, and are consequently not estimated. Also, owing to the relatively high viscosities of dispersions, it is often difficult to take samples with a syringe. For the same reasons, there are frequently problems associated with the addition of internal standards to such samples.

A variety of methods is available for sampling aqueous dispersions of polymers, and a few of them will be mentioned here.

(*a*) Direct sampling of the dispersion with a syringe, if necessary, after diluting the dispersion with water or a mixture of water and a water-soluble organic solvent.

(*b*) Preparation of a clear solution of the dispersion in a suitable solvent. The solvent used may be miscible with water, or quantitatively react with it.

(*c*) Isolation of the residual monomers by steam distillation or azeotropic distillation.

(*d*) Removal of the monomers and all the volatile components from the dispersion by means of the carrier gas. The dispersion is contained in a precolumn directly connected to the gas chromatographic column.

(*e*) Application of head space analysis (*see* 2.15).

For routine analysis, direct sampling (method *a*) appears at first sight to be the simplest and fastest procedure, since the FID does not respond to water and the polymer will remain behind in the injection unit. However, it is then essential that the gas chromatograph be fitted with an interchange-able glass insert, which must be renewed after each analysis. Furthermore, the volatilization temperature must be carefully chosen, since if it is too low there is a risk that not all of the monomers will be quantitatively determined. On the other hand, if the temperature is too high degradation products can be formed from the polymer, giving a misleading value for the residual mono-mer content. It is often impossible to sample the highly viscous or gelatinous dispersion and thus they have to be diluted. For example, after diluting a dispersions with water in the ratio 1:1, Shapras and Claver determined 45 ppm acrylonitrile, 0·12% styrene and 0·67% 2-ethylhexyl acrylate (Fig. 164).[20]

Since errors can occur in this dilution process as a result of demulsification, it is advisable to use mixtures of water with acetone of dioxan in the ratios 80:20 and 50:50 respectively, instead of pure water. Obviously the chromato-graphic peak of the organic solvent chosen must not overlap those of the residual monomers being analysed.

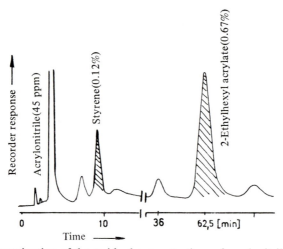

Fig. 164   Determination of the residual concentrations of acrylonitrile, styrene and 2-ethylhexyl acrylate in an aqueous polymer dispersion (column: 1·8 m 10% stearic acid amide of propyldimethyl-β-hydroxyethylammonium nitrate on Chromosorb W, column temperature: 100°C) (from P. Shapras and G. C. Claver, *Anal. Chem.*, **34**, 433 (1962)).

Wilkinson *et al.* have measured residual monomers by dissolving the dispersion in appropriate solvents (method b).[21] Acetic anhydride, 2,2-dimethoxypropane and cyclohexanone were reported to be suitable solvents for the most common dispersions. The first two compounds have the advan-tage in that they remove the water by chemical reaction to form acetic acid, or acetone and methanol. This technique is to be preferred when a thermal conductivity detector is being used. The appropriate solvents for particular dispersions are listed in Table 11.

Wilkinson *et al.* recommend polyethylene glycol 6000 as the stationary phase for the analysis of residual monomers, and the retention volumes on this stationary phase at appropriate temperatures are summarized in Table 12.

Figure 165 shows the chromatogram of an acrylonitrile–butyl acrylate dispersion dissolved in acetic anhydride.

TABLE 11 Solvents for various polymer dispersions
(from L. B. Wilkinson *et al., Anal. Chem.,* **36,** 1759 (1964))

Type of dispersion	Solvent
Vinyl chloride–vinylidene chloride	cyclohexanone
Vinyl acetate–vinyl propionate	cyclohexanone
Butadiene–styrene	cyclohexanone, dimethoxypropane
Styrene–2-ethylhexyl acrylate	dimethoxypropane
Styrene–2-ethylhexyl acrylate–acrylonitrile	dimethoxypropane, acetic anhydride
Ethyl acrylate–butyl acrylate–styrene	dimethoxypropane, acetic anhydride
Ethyl acrylate–methyl methacrylate	acetic anhydride
Acrylonitrile–butyl acrylate	acetic anhydride

It should be noted that the reaction between acetic anhydride and water must be catalysed with trichloroacetic acid. A similar requirement applies to the corresponding reaction between water and 2,2-dimethoxypropane, where methanesulphonic acid acts as a catalyst. Figure 166 illustrates the determination of 2-ethylhexyl acrylate and styrene by dilution of an aqueous dispersion with 2,2-dimethoxypropane.

The measurement of residual vinylidene chloride and vinyl chloride may be cited as an example of the use of cyclohexanone as a solvent for a dispersion, although a slight turbidity persists in the solution (Fig. 167).

Polymer dispersions may also be dissolved in methanol or dimethyl-formamide for analysis of their residual monomer content. Figure 168 shows the determination of vinyl acetate, n-butyl acrylate and 2-ethylhexyl acrylate in a polyvinyl acetate dispersion dissolved in methanol.[22]

TABLE 12 Retention volumes of some monomers on polyethylene glycol 6000
(from L. B. Wilkinson *et al., Anal. Chem.,* **36,** 1759 (1964))

Monomer	Column temperature (°C)	Net retention volume (ml)
Methyl acrylate	150	238
Methyl methacrylate	150	272
Ethyl acrylate	150	280
Isobutyl acrylate	150	500
n-Butyl acrylate	150	663
2-Ethylhexyl acrylate	200	865
Styrene	200	430
1,3-Butadiene	100	56
Vinyl chloride	150	45
Vinylidene chloride	150	90
Vinyl acetate	150	193
Vinyl propionate	150	272
Acrylonitrile	150	308

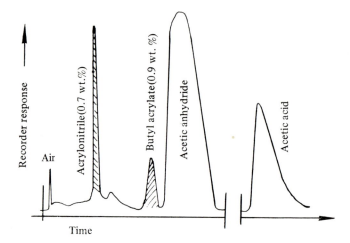

Fig. 165   Determination of residual monomers in an acrylonitrile-butyl acrylate dispersion dissolved in acetic anhydride (column: 3 m polyethylene glycol 6000 on Chromosorb, temperature: 150°C) (from L. B. Wilkinson *et al., Anal. Chem.,* **36,** 1759 (1964)).

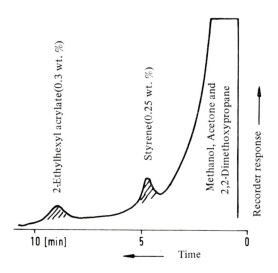

Fig. 166   Determination of residual styrene and 2-ethylhexyl acrylate in an aqueous polymer dispersion dissolved in dimethoxypropane (column: 3 m polyethylene glycol 6000 on Chromosorb, column temperature: 200°C) (from L. B. Wilkinson *et al., Anal. Chem.,* **36,** 1761 (1964)).

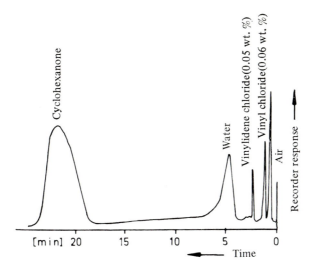

Fig. 167 Determination of residual vinyl chloride and vinylidene chloride in an aqueous polymer dispersion dissolved in cyclohexanone (column: 3 m polyethylene glycol 6000 on Chromosorb, temperature: 150°C) (from L. B. Wilkinson *et al.*, *Anal. Chem.*, **36**, 1761 (1964)).

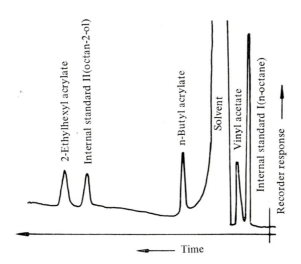

Fig. 168 Determination of residual monomers in a polyvinyl acetate dispersion dissolved in methanol (column: 2 m Carbowax 1500, temperature programme: 50–110°C) (from G. N. Fossick and A. J. Tompsett, *J. Oil Colour Chemists Assoc.*, **49**, 477 (1966)).

Another method of measuring residual monomers in aqueous dispersions involves separation by distillation with toluene, which at the same time can act as an internal standard (method c).[23] This procedure can be used to analyse, e.g., ethyl acrylate and also acrylonitrile, although in the latter case there is a small error owing to the slight solubility of the compound in water. The method can only be used for those monomers which can be completely extracted from water with toluene.

Nelsen et al. have described a further possible method of measuring residual monomers, which avoids the use of a syringe.[24] The aqueous dispersion is weighed in a glass capillary tube, and the components to be analysed are flushed into a short chromatographic column by the carrier gas. After being separated in this way, the volatile materials are oxidized over copper oxide to water and carbon dioxide. The water is removed with calcium sulphate, and the carbon dioxide is then measured with a thermal conductivity detector. When this method is employed for styrene in styrene–butadiene copolymers, it has the advantage that response factors do not have to be taken into account.

Head space analysis (method e) is certainly the best means of determining residual monomers and other volatile compounds in aqueous dispersions of polymers, or in polymer dispersions in general, since it largely avoids the difficulties mentioned for methods a to d (see 2.15).

ANALYSIS OF PLASTICIZERS AND OTHER ADDITIVES

In addition to monomers and other low molecular weight compounds which are present as a result of the manufacturing processes used for synthetic polymers, there are also a number of other materials which are added to improve their properties, and it is therefore necessary to be able to analyse for these.

Plasticizers are an example of this group of substances. Their migration into foodstuffs presents a considerable problem, since not only do they themselves diffuse, but they also act as carriers for other compounds which are present in the packaging material.[25] Although thin-layer chromatography provides an excellent means of analysing these substances down to certain concentration levels, because of the inherent sensitivity of the detectors, gas chromatography is also employed for the quantitative analysis of these very high boiling point compounds, some of which are themselves utilized as stationary phases.[26–29] Thus, gas chromatography provides the only method of determining 2 ppm of diethyl phthalate in foodstuffs.[30] Cook et al. have used a 23 cm long column having silicone grease as the stationary phase for the determination of dibutyl phthalate and dibenzyl phthalate at 235°C.[31] Zulaica and Guiochon have utilized a pyrolysis unit for the determination of plasticizers.[32] Care must be taken to ensure that the pyrolysis temperature is not too low, otherwise the plasticizer would not be completely vaporized. If the temperature is too high there is the risk of decomposition. The optimum conditions are reported to be a volatilization temperature of 625°C for 10 seconds duration. The products resulting from this procedure were resolved on two columns. The first was 2 m long and packed with 0·5 % silicone rubber SE 30 on 125–160 μm glass beads. The second column was 3 m long and contained glass beads which had been lightly etched with hydrofluoric acid and coated with 0·5 % polyneopentyl glycol adipate. The column temperature was between 200 and 240°C. Analysis of three plasticizers (tri-n-butyl phosphate, di-n-butyl phthalate and di-n-butyl sebacate) in PVC was used to demonstrate the practicability of this vaporization technique, and to compare it with the conventional sampling of solutions of the plasticizers. It was found that 10 % decomposition must be allowed for in the vaporization method.

Another possible technique is to extract the polymer with suitable solvents and carry out gas chromatographic analysis of the plasticizers in the extract. For example, Diemair and Pfeilsticker have used nitromethane in a determination of plasticizers which had diffused from polymers into foodstuffs.[25]

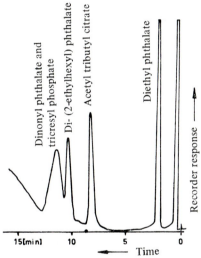

Fig. 169   Analysis of plasticizers on a 30 cm column of 3 % Silopren on Chromosorb R, temperature programmed from 160 to 500°C (from W. Diemair and K. Pfeilsticker, *Z. Anal. Chem.*, **212**, 53 (1965)).

A 30 cm column of 3 % Silopren on Chromosorb R was temperature programmed from 160°C and an FID was employed. This method may be used to determine plasticisers having boiling points up to about 500°C. Figure 169 shows a typical chromatogram for 8 μg each of diethyl phthalate, di-(2-ethylhexyl) phthalate, acetyl tributyl citrate, dinonyl phthalate and tricresyl phosphate.

Haase has used ether to extract the plasticizers from artificial leather, and separated them gas chromatographically on silicone oil and silicone grease stationary phases.[33] They were identified by measuring their retention times, and also by infrared spectroscopy. The chromatogram of an ether extract of a sample of artificial leather is shown in Fig. 170.

Plasticizers may also be isolated by dissolving and reprecipitating the polymer, and the solution can then be analysed by gas chromatography. Esposito has used this procedure to analyse nitrocellulose, vinyl and acrylic paints.[34] They were diluted with acetone and precipitated with carbon tetrachloride or petroleum ether, and the resulting plasticizer solutions

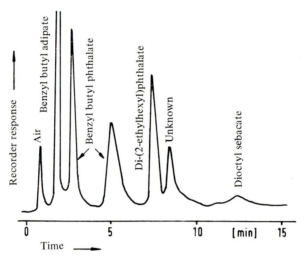

Fig. 170   Gas chromatographic analysis of an ether extract of a sample of artificial leather (column: 2 m silicone grease at 296°C) (from H. Haase, *Kautschuk Gummi*, **20**, 506 (1967)).

concentrated by evaporation. The concentrates were analysed on a column of silicone grease on Chromosorb W, temperature programmed from 210 to 290°C at 4 deg C/min. Dibutyl sebacate and dibutyl adipate were used as internal standards. Figure 171 illustrates the determination of diethyl phthalate (*A*) and dibutyl phthalate (*B*) in a nitrocellulose paint with dibutyl sebacate (*C*) as the internal standard.

Another group of materials added to polymers are the substituted phenols, *i.e.* antioxidants, which inhibit their thermal oxidation. These substances can be isolated by extraction, *e.g.* from polyethylene with carbon disulphide, iso-octane, carbon tetrachloride or chloroform.[35] Schröder and Rudolph have determined 2,6-di-*t*-butyl-4-methylphenol after extraction with chloroform.[36] Jennings *et al.* used silicone grease and polypropylene glycol as stationary phases for the separation of numerous phenolic antioxidants, and showed that 2,6-di-*t*-butyl-4-methylphenol can be quantitatively analysed in packaging materials with an error of $\pm 3\%$.[37]

Long and Guvernator have described the use of hexane to extract this antioxidant from polyethylene, followed by measurement with an electron capture detector, which is relatively insensitive towards the solvent (hexane).[38]

Stabilizers and ultraviolet-absorbers in synthetic polymers may also be determined by gas chromatography, after extraction with hexane or carbon disulphide.[39,40] Polyolefins also contain antistatic additives such as alkyl-diethanolamines and alkyldiethanolamide which can also be measured by gas chromatography after extraction and conversion to the trimethylsilyl ethers.[41]

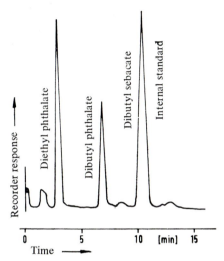

Fig. 171   Analysis of plasticisers on a silicone grease column temperature programmed from 210 to 290°C at 4°C/min (from G. G. Esposito, *Anal. Chem.*, **35**, 1440 (1963)).

References
1. PFAB, W. and NOFFZ, D., *Z. Anal. Chem.*, **195**, 37 (1963).
2. EISENBRAND, J. and EICH, H. W., *Z. Anal. Chem.*, **175**, 4 (1960).
3. KLESHCHEVA, M. S., BALANDINA, V. A., USACHEVA, V. T. and KOROLEVA, L. B., *Vysokomolekul. Soedin*, Ser. A, **11**, 2595 (1969).
4. SHAPRAS, P. and CLAVER, G. C., *Anal. Chem.*, **36**, 2282 (1964).
5. SIMPSON, D., *Brit. Plast.*, **41** (5), 78 (1968).
6. EGGERTSEN, F. T. and STROSS, F. H., *J. Appl. Polymer Sci.*, **10**, 1171 (1966).
7. CROMPTON, T. R. and MYERS, L. W., *Plastics and Polymers*, 205 (1968).
8. PARKHURST, R. M., RODIN, J. O. and SILVERSTEIN, R. M., *J. Org. Chem.*, **28**, 120 (1963).
9. BRAUN, D. and MEIER, W., *Angew. Makromolekulare Chem.*, **1**, 17 (1967).
10. STEIN, D. J. and MOSTHAF, H., *Angew. Makromolekulare Chem.*, **2**, 39 (1968).
11. KURZE, J., STEIN, D. J., SIMAK, P. and KAISER, R., *Angew. Makromolekulare Chem.*, **12**, 25 (1970).
12. STEVENS, M. P. and PERCIVAL, D. F., *Anal. Chem.*, **36**, 1023 (1964).

13. STEVENS, M. P., *Anal. Chem.*, **37**, 166 (1965).
14. LEONARD, R. E. and KIEFER, J. E., *J. Gas Chromatog.*, **4**, 142 (1966).
15. LANGER, S. H., PANTAGES, P. and WENDER, I., *Chem. Ind.* (London), 1664 (1958).
16. SHULGIN, A. T., *Anal. Chem.*, **36**, 920 (1964).
17. ONGEMACH, G. C. and MOODY, A. C., *Anal. Chem.*, **39**, 1005 (1967).
18. ZILIO-GRANDI, F., SASSU, G. M. and CALLEGARO, P., *Anal. Chem.*, **41**, 1847 (1969).
19. MORI, S., FURUSAWA, M. and TAKEUCHI, T., *Anal. Chem.*, **42**, 661 (1970).
20. SHAPRAS, P. and CLAVER, G. C., *Anal. Chem.*, **34**, 433 (1962).
21. WILKINSON, L. B., NORMAN, C. W. and BUETTNER, J. P., *Anal. Chem.*, **36**, 1759 (1964).
22. FOSSICK, G. N. and TOMPSETT, A. J., *J. Oil Colour Chemists Assoc.*, **49**, 477 (1966).
23. TWEET, O. and MILLER, W. K., *Anal. Chem.*, **35**, 852 (1963).
24. NELSEN, F. M., EGGERTSEN, F. T. and HOLST, J. J., *Anal. Chem.*, **33**, 1150 (1961).
25. DIEMAIR, W. and PFEILSTICKER, K. Jr, *Z. Anal. Chem.*, **212**, 53 (1965).
26. PEEREBOOM, J. W. C., *Chem. Weekblad.*, **57**, 625 (1961).
27. PEEREBOOM, J. W. C., *J. Chromatog.*, **4**, 323 (1960).
28. BRAUN, D., *Kunstoffe*, **52**, 2 (1962).
29. DIEMAIR, W. and GÖCKEL, J., Dissertation, University of Frankfurt on Main, 1962.
30. WANDEL, M. and TENGLER, H., *Deut. Lebensm. Rundschau*, **59**, 326 (1963).
31. COOK, C. D., ELGOOD, E. J., SHAW, G. C. and SOLOMON, D. H., *Anal. Chem.*, **34**, 1177 (1962).
32. ZULAICA, J. and GUIOCHON, G., *Anal. Chem.*, **35**, 1724 (1963).
33. HAASE, H., *Kautschuk Gummi*, **20**, 501 (1967).
34. ESPOSITO, G. G., *Anal. Chem.*, **35**, 1439 (1963).
35. SPELL, H. L. and EDDY, R. D., *Anal. Chem.*, **32**, 1811 (1960).
36. SHRÖDER, E. and RUDOLPH, G., *Plaste Kautschuk*, **10**, 22 (1963).
37. JENNINGS, E. C., CURRAN, T. D. and EDWARDS, D. G., *Anal. Chem.*, **30**, 1946 (1958).
38. LONG, R. D. and GUVERNATOR, G. C., *Anal. Chem.*, **39**, 1493 (1967).
39. LAPPIN, G. R. and ZANNUCCI, J. S., *Anal. Chem.*, **41**, 2076 (1969).
40. ROBERTS, C. B. and SWANK, J. D., *Anal. Chem.*, **36**, 271 (1964).
41. DAVIES, J. T. and DENHAM, B. H., *Analyst*, **93**, 336 (1968).

INDEX